Making DVDs

Making DVDs

A Practical Guide to Creating and Authoring Your Own Discs

Lee Purcell

McGraw-Hill

New York Chicago San Francisco Lisbon London Madrid
Mexico City Milan New Delhi San Juan Seoul
Singapore Sydney Toronto

Cataloging-in-Publication Data is on file with the Library of Congress

McGraw-Hill

A Division of The McGraw-Hill Companies

1 2 3 4 5 6 7 8 9 0 DOC/DOC 0 1 0 9 8 7 6 5 4

P/N 143705-3
PART OF
ISBN 0-07-143191-8

The sponsoring editor for this book was Stephen S. Chapman and the production supervisor was Pamela A. Pelton. It was set in Giovanni Book by Lightspeed Publishing LLC. The art director for the cover was Anthony Landi.

Printed and bound by R. R. Donnelley.

This book is printed on recycled, acid-free paper containing a minimum of 50% recycled, de-inked fiber.

To Jordan Hemphill for guiding me through
my first years as a technical writer.

Table of Contents

1

Introduction: The Promise of DVD Technology

Communicating with light—the idea has a wonderful elegance and a poetic simplicity. The Digital Versatile Disc, the DVD, channels light into coherent symbols: words, moving pictures, sound. The concept is not so much revolutionary as evolutionary, arising from the success of the audio compact disc, which in turn gave form to the CD-ROM. The planners of the five DVD books took the best ideas distilled from more than a dozen years of evolving CD-ROM standards and designed a framework and delivery platform tailored to a new generation of digital media.

The medium of exchange is binary, collections of ones and zeroes combined in innumerable ways to represent a broad palette of knowledge and experience. A single DVD disc can contain hundreds of thousands of pages of words, hours of video content, rich music that significantly surpasses the audio capabilities of the compact disc, or virtually anything that can be encoded in digital form. The result is a medium that changes the nature of the way we communicate.

Coupled with the increasing capabilities of personal computers, which now rival the supercomputers of years past, anyone with an idea and a modest amount of equipment can create a DVD. This simple fact opens up abundant possibilities for developers, producers, small businesses, artists, musicians, and others who enjoy creating digital media. Independent filmmakers now have a venue that is accessible and affordable. Videographers and documentary producers can shape their work to take advantage of the distribution flexibility offered by a quarter-ounce disc. Photographers can organize and display works on disc, playable on computers or televisions or portable DVD players. Musicians can compose for the extra spectral dimensionality of five-channel audio sampled at very high resolutions. Animators can design works easily delivered to a broad audience. Corporate presentations can take on the extra vibrancy and immediacy offered by digital video and surround sound. Technical communicators, marketing professionals, corporate trainers, motivational speakers, educators, and many others can benefit from the communication avenues offered by DVD. The tools of creation can be mastered by anyone with a facility for working with computers and the diligence to investigate and master the craft. New applications for DVDs will continue to arise as experimenters and designers and creators and artists of all types recognize and embrace the potential of this medium.

Expanding the Scope of Human Communication

Before Gutenberg devised the idea of movable type to dramatically expand the possibilities of typography, human communication depended largely on face-to-face communication. While pictographs, hand-lettered scrolls, block-printed pages, and other methods of conveying ideas through symbols have existed for thousands of years, widespread, portable methods for communicating were clearly lacking. Books, hand lettered by scribes or printed laboriously using wood blocks or clay characters, typically belonged only to a limited segment of society—scholars, rich men, civil authorities, or monks.

Refinements in printing technology made it possible to inexpensively create books, newspapers, and pamphlets, leading to a vastly different model for distributing information and ideas. Printing played a crucial role in the birth of the American Revolution as ideas communicated by patriots such as Benjamin Franklin and Isaiah Thomas helped shape and galvanize the move towards independence. Books and printed materials of all types flourished worldwide as processes and techniques were improved, making it easier to inexpensively reach large audiences. These processes, though, still depended on the availability of a fairly expensive piece of equipment: the printing press. Clearly, not everyone could afford their own personal printing press. Publishers were established to channel the work of writers through the production and distribution processes. The end-to-end model involved large numbers of individuals to get the words of the writer onto the printed page and into the hands of readers.

The twentieth century brought new communication technologies to the forefront—photography, audio recording, motion pictures, television, and radio opened up a broader spectrum of communication possibilities, but each of these forms of media required significant investments in equipment and technology, resulting in industries that again relied on elaborate production methods and the coordinated efforts of many different individuals to complete a feature film, phonograph record, or television broadcast.

With the DVD, the tools for radical new forms of communication are in the hands of the creators. Digital media is malleable and flexible within the confines of a desktop computer. The software applications that capture and process video, audio, and graphics take advantage of monumental increases in processing speed that make it feasible to manipulate millions of bits of digital content in real time. Sophisticated compression techniques make it possible to encode and decode streams of audio and video so that they can be delivered seamlessly on playback devices and computer systems.

The development of DVD recorders and their subsequent price reductions over the last few years completed the picture. The development is as striking as if printing presses were so affordable in the 18th century that every family could own one. Cost-effective DVD recorders brought the final element home, completing the digital content production stream, from capture to processing to recording. The availability of inexpensive DVD duplicators extends the possibilities even further, letting developers and producers generate high-quality, printed DVD discs that can be played in the majority of players in the market.

The unprecedented degree of creative freedom afforded by the personal computer surpasses many of the media capabilities we have fostered and refined over the last century. Motion pictures and still photography rely on chemically dependent processes that pollute air and water. Digital video and digital photography take advantage of inexpensive storage techniques that provide an immediacy and accessibility not possible with film. The complexities and expenses associated with analog recording studios can be bypassed in favor of digital recording methods and software applications that can virtually simulate any number of sonic landscapes. Animation and video effects can be accomplished on any current generation personal computer with results that rival high-end dedicated workstations of a few years ago. Budding film creators such as Steven Spielberg and Steven Soderbergh may have gotten their starts with 8mm and 16mm cameras, but the next generation of creative filmmakers will have far more sophisticated options through digital video camcorders, non-linear editing tools, and DVD recorders.

In truth, the capabilities of the latest digital tools and the DVD range far beyond the efforts of the hobbyist and casual user. For a wide variety of applications, the desktop computer platform can produce results of startling audio and video clarity. While small-scale independently produced DVDs can't yet directly compete with multimillion-dollar Hollywood epics, they can run rings around Hollywood in providing intelligent alternative entertainment, education, and original ground-breaking material. As most readers are also probably aware, the mainstream filmmaking community is increasingly turning to digital production techniques for both animated 3-D works (such as Pixar's *A Bug's Life*) and feature films (such as Boyle's *28 Days Later*). New distribution and delivery techniques for digital video are also being introduced, promising a far less expensive means of copying and distributing films than is the case with heavy 35mm projection reels. DVD discs feature prominently in this new approach to film distribution.

Learning By Doing

By design, this book emphasizes learning by doing and practical applications over theory, specifications, and pedantry. The information contained here is intended to serve as a springboard for those who want to test the medium and use it as a form of expression. The technology surrounding DVD continues to change at rapid rates. DV camcorders, authoring software, computer platforms, and standards also progress and change with breathtaking rapidity. Although many of the examples and discussions included in this book mention specific equipment and software products, the points made in these discussion should transcend the limited lifespans of feature sets and product capabilities. Examples provide guidelines and offer perspective, rather than dictating a fixed course of action.

Exactly how do you make a DVD? The equipment requirements are fairly minimal, as are the software requirements, described in Chapter 2. You can create a compelling piece of digital media content and burn it to DVD without even owning (or renting) a camcorder. Using stills or animation, music and dialog, you can tell a story, teach a skill, or produce a work of audio-visual art. The temptation may be to race out and buy a $5000 High-Definition DV camcorder, a $3000 non-linear editing suite, and the fastest computer in the store. But, I'd suggest starting more modestly. Take twelve of your favorite photographs, scan them into your computer, manipulate them in Adobe Photoshop, and then see what kind of two-minute short you can create in your favorite video editing application. Burn the results to DVD and see how it looks when displayed on a television screen or on a computer display. One of the most creative filmmakers in modern times, Stan Brakhage, produced some of his most compelling films by painting directly on 35mm leader, frame by frame. His work and techniques are discussed in Chapter 11.

Another useful way to learn a process is to watch someone else do it. In that spirit, a number of case studies are presented in the latter chapters of this book, each one illustrating how an individual or project team developed a particular type of DVD. These case studies cover projects by some of the leading companies working to produce DVDs, as well as individuals who are creating innovative, homegrown projects. The lessons contained therein illustrate both the pitfalls of making DVDs, as well as the techniques that can help you gain the best results.

If you are starting from ground zero and would like to get oriented to the technical aspects of the DVD formats, Chapter 17, *DVD Fundamentals,* gives a selective overview of the most important elements of DVDs. This information can help you make decisions about selecting media, choosing duplication or replication, and determining the appropriate DVD format for a project.

The most important thing is to get started and make a DVD. The tools aren't expensive and the techniques aren't difficult to learn. Some of the best schools are those that go easy on the theory and concentrate on gaining real-world experience. At New York University, where film is a major part of the curriculum, students are turned loose with a camera in New York city and told to make a film. From the many successful independent filmmakers who trace their roots to NYU, the process appears to work.

The first few chapters of this book provide guidelines and address the most common questions that arise when someone tackles some aspect of DVD creation, whether concerning the equipment required, the options for handling content, or the ultimate replication and distribution of the completed title. These chapters include abundant pointers to other sources of information—the goal is to steer the reader through the thicket of extraneous information and overwhelming technical details to reach a clearing where the DVD creation and production process becomes understandable. It is not as important which platform you perform the authoring on or what equipment you use to capture the video or what styles and genre you prefer to work in. The measure of a quality DVD project is the end result displayed before an appreciative audience.

Entering New Territory

In many ways, the DVD is enlarging the spectrum of communication and entertainment. The inexpensive replication and shipping costs of the medium, coupled with the interactivity that can layer content (such as director's commentary) and provide a more flexible user experience, has led to a wide range of innovative uses. Developers alert to market opportunities and potential may be able to use these ideas as seeds for new projects.

Renaissance in Film Appreciation

DVD has changed the way that films are watched, bringing them into the home in their original formats (through letterboxing) with audio that sounds great on modest speakers as well as the most expensive surround sound systems. Suddenly, the viewing public has gained an interest in the history, background, production, and esoterica surrounding film. The spark of interest that was ignited in movie connoisseurs with the Laserdisc has spread to a full-blown prairie fire with the DVD. Companies such as the Criterion Collection, as discussed in Chapter 11, are revitalizing titles that enjoyed modest success on Laserdisc or VHS and finding new, appreciate audiences on DVD.

Audio commentary by directors, actors, writers, and other participants in the film creation process has opened the door to the public, providing a behind the scenes view of how the magic of Hollywood is accomplished. To some degree, you might think that knowing how the tricks are executed in film (like how those *Jurassic Park* dinosaurs can pick up characters and fling them around) would dampen the entertainment enjoyment, but the appetite for commentary seems insatiable. Listening to Salma Hayek discuss her role in *Frida* or Martin Scorsese offer insights into *Taxi Driver* can be just as compelling as the movie itself. The fact that audio commentaries are generally considered an essential component of a DVD feature film has led to much of this work being hastily prepared. At its best, this feature can have all the depth and insight of attending a film school class taught by a very good instructor. At its worst, the commentary can be mere filler on the DVD. As a DVD developer, take care to use commentaries as an enhancement to the content rather than a meaningless, vacuous extra soundtrack.

Moving Television Shows to DVD

Even the best television shows suffer from disjointed commercial interruptions, with stories neatly segmented to allow breaks for pointed sales messages. Sometimes the best way to view these shows is on DVD and, increasingly, producers are bundled whole seasons on DVD and releasing them as a package—without commercial interruptions. Shows such as *Sex and the City, West Wing, Six Feet Under*, and *Alias* have been released in DVD format, often containing many extras to please fans, taking them behind the scenes for intimate conversations with the stars.

The medium also offers an outlet for television shows that may have been prematurely cancelled, despite the fact they may have been well-produced entertaining works that never found their audience. At the other end of the spectrum are obscure, unusual or incredibly bad television shows that can be fun to watch just to see how bad television programming can get.

Often, television shows on DVD contain commentary tracks. Revealing little glimpses of production decisions that are worked into commentary for shows like *The Sopranos* let fans gain a sense of peering into the director's or writer's brain, which is often an entertaining place to be.

Not every show on television is worth watching over and over, but many of the shows that make it to DVD can provide entertainment and insights and good viewing when there is nothing else of value on broadcast TV. Once again, DVD has extended an existing medium, adding convenience, extra dimension, and longevity.

Theatrical Performances on DVD

Exacting playwrights such as Edward Albee sometimes look at film adaptations as being a distortion of their work. But, given some creative control, they often welcome the opportunity to communicate in a new venue. One of Albee's plays, *The Delicate Balance*, was released on DVD featuring an interview with the playwright. He sees maintaining the integrity of the work as being crucial to any translation to film or video, and strongly states that not a word of text should be altered to accommodate the medium.

Plays are finding their way to DVD in increasing numbers, some from videotaped or filmed captures of the stage performances, others from performances originally broadcast on television. The tempo and pacing of a typical play differ substantially from a feature film, requiring some adjustment of audiences unaccustomed to long uninterrupted sequences of dialog. But, for those who enjoy this engaging art form, the DVD is becoming a channel to revisit classic works (such as O'Neill's *Iceman*

Cometh and Chekhov's *Three Sisters*, released by Kino) that were originally produced on film.

Early television broadcasts dating back as far as the 1950's have also found a new market through DVDs. Broadway Theater Archive has laboriously restored a wide range of works, including the Arthur Miller masterpiece, *Death of a Salesman*, first broadcast in 1966. Distributed through Kultur Video (*www.kultur.com*), these early works, many of them from the golden years of public television, have reached a whole new audience in their optical disc reincarnations.

A Growing Audience

The audience for any prospective DVD title is large and growing. One of the most compelling aspects of working with DVDs is the abundance of playback options and the wide distribution of players. At least for the present, you don't have to worry about creating a DVD title for a medium that will be obsolete in two years. An enormous amount of industry momentum and energy back DVD technology at this point and the oft-repeated claim that DVD is the fastest growing consumer entertainment technology of all time is true.

During 2003, sales of DVD set-top players are expected to exceed 66-million units, contributing to a total installed based of over 200-million players. This, of course, does not include the totals for all of the computer systems equipped with DVD-Video playback capabilities, as well as the new generation of DVD recorders that can record broadcast television programs, as well as playback commercial DVD-Video discs.

With a potential audience of this magnitude, the market is poised for new, intelligent DVD titles crafted by a new generation of producers. The tools are available. The market is open. If you have an idea for a unique title, now is the time to launch a DVD project.

Steps Involved in Making a DVD

The individual processes that are involved in making DVDs vary from project to project, but the following steps are fairly typical.

- **Acquiring Assets**: The content that you put on a DVD can come from a wide variety of sources. Depending on the nature of a project, the video content can originate from film (converted to digital format through a telecine process), analog video (converted to digital format through conversion) or digital video (which can consist of DV, DigiBeta, HD digital video, or other formats). The pre-production phase of a DVD project may include purchasing or licensing assets, such as music, still photographs, existing film or video footage, and so on.
- **Production**: The production phase of a DVD project differs depending on whether the content is a video feature (designed for a DVD-Video presentation), an animated work (which could be included in either DVD-Video or a DVD-ROM title), or a multimedia presentation designed for use on a DVD-ROM. If you are capturing original video material for a project, the production phase includes all the work involved in videotaping content, including the lighting, sound recording, scene preparation, and so on. At the end of the production phase, you should have a full complement of assets ready to turn into a polished production.
- **Post-Production**: Working from a library of collected assets, post-production involves the work that it takes to edit, combine, and enhance assets. Often, for a typical DVD project, this is the work that takes place in a non-linear editing application, such as Adobe Premiere, Sonic Foundry Vegas, or Apple Final Cut Pro. The post-production phase of a project often takes far longer than the production phase, particularly for feature films, long documentaries, or other works that exceed an hour or so in length. During post-production, music is sometimes added to soundtracks, sound is processed to improve quality, special effects and transitions are added to video content, and so on. At the conclusion of the post-production stage, a developer should have a full set of files in formats that can be used in a DVD authoring application.
- **DVD Authoring**: During the authoring stage, the developer organizes all of the assets into a specific format required for presentation on a DVD player. This includes designing menus to present the content, and categorizing media files into four basic groups,

including audio, video, subpictures, and still pictures. Within the authoring application, you create the global properties that apply to a title, such as regional coding, PAL/NTSC video, copy protection, and similar properties. The order of the video files is assembled for playback, usually on some form of timeline, and entry points to different sections of video can be handled by menu buttons. You essentially create a navigation system that provides access to the content on the disc. It can be as simple as playing back a 60-minute video from start to finish, or as complex as a dozen menu and submenus that provide access to all of the different materials on disc.

- **Pre-Mastering**: Pre-mastering includes all of the final processes to prepare content for submission to a replication service or to create a master for duplication. Many times these processes are set up and handled by the DVD authoring application. The pre-mastering process combines all of the audio and video content, as well as the subpictures, into compiled, multiplexed VOB files that will be stored in specific folders on the DVD disc. This content is organized to meet the requirements of a replication service, outputted to DLT or DVD-R authoring use media. It can also be burned to DVD-R general use media in a form that allows playback in the majority of DVD players. DVD-R general use discs are often used as masters in DVD duplicators for runs that are typically range from ten to several hundred discs.
- **Replication or duplication**: If you are producing a commercial disc for mass distribution, a replication service takes your DLT or DVD-R authoring media and prepares a glass master, which is used to produce a series of stampers used in the manufacturing process. During the replication phase, disc labels and packaging materials are also created. Disc surfaces can be silk-screened, printed through lithographic processes, or imprinted in other ways. Packaging materials can range from a simple fiber sleeve to the conventional packaging, such as the safe box, found in the racks at retail outlets. For projects designed for DVD duplication, labels can be added through disc printing, preprinting, or adhesive labels added after duplication. Packaging for duplicated discs can be as simple or as elaborate as commercially manufactured discs.

The following chapters explain the specific details on how these different processes are accomplished.

2

Getting Equipped: From Camcorders to DVD Duplicators

What do you need to get started? The range of equipment for making DVDs can be incredibly complex, but the essential equipment and software needed to successfully create DVD content and burn it to a disc is within the reach of most computer users. This chapter examines the equipment options that can get you started.

Selecting a Computer Platform

For most purposes, a computer system will serve as center of your DVD production work. An effective computer system should have a sufficiently fast processor to handle demanding video editing work, enough system memory to keep two or three applications running while working, a decent amount of hard disk storage (at least 80GB), and a high-speed interface (most frequently, an IEEE-1394 or FireWire connection) to connect your DV camcorder to the system for transferring the video. Alternately, if you need to bring in analog video material into the system (for example, if you have a Sony BetaCam SP camcorder that you want to use for video work), you need a video capture board. This acts as a high-speed, high-quality interface to convert the analog material to digital format for use in an editing suite. Whenever possible, however, keep all your tools—camcorder, non-linear editor, output device—in the digital realm. This avoids generational and transfer losses, which for high-end formats such as BetaCam SP is not extreme. Overall, however, your results will be more controllable and more pleasing if you don't need to cross the analog and digital divide two or three times.

Minimal Windows Platform for Getting Started

As I began work on this book, I began hunting for an external DVD recorder to use with an existing Microsoft Windows machine. Prices ranged from around $350 to $500 for a capable recorder. At the time, Dell Computer Corporation was offering a special that included an internal DVD+R recorder and the bundle was irresistible. For around $550, I obtained what I consider a practical baseline machine for making DVDs. The Dell Dimension DIM 2350 included:

- Intel Pentium 4 processor, running at 2.00GHz with 256MB of system RAM
- NEC DVD+RW ND-1100A, which capably handles DVD+R, DVD+RW, CD-R, and CD-RW recordable media
- 30GB Hard disk drive

To this package, I added two additional components:

- An IEEE-1394 interface board with three I/O connectors
- A 120GB MicroNet Advantage hard disk drive with IEEE-1394 interface (the standard Dell bundled drive was too small for anything other than short video editing projects)

This system has worked flawlessly with a variety of NLE applications and authoring programs. The editing/authoring software included on the system, Sonic MyDVD!, features very rudimentary capabilities, but you can very quickly edit a few video clips and burn them to DVD to test your components. To get into more serious DVD title creation, you should upgrade to a mid-level editing/authoring package. Three options include:

- **Sonic ReelDVD**: Designed for independent and corporate video projects with a price tag around $700. More details at *www.sonic.com/products/reeldvd/default.asp.*
- **Sony Vegas+DVD**: Integrates video editing and DVD authoring with a price of about $700. More details at *mediasoftware.sonypictures.com/products/showproduct.asp?PID=810.*
- **Avid Xpress Pro**: Supports many professional features, including HD editing and 24p input and output. Priced around $1700. Details at *www.avid.com/products/xpresspro/index.asp.*

Minimal Macintosh Platform for Getting Started

Apple took an early lead in the DVD development arena by including inexpensive DVD-R drives in select systems, making it possible for independent and corporate developers to burn DVDs for projects. At the time, the mainstream option for serious DVD recording was a Panasonic machine priced at over $5000. A subtle difference separated the $5000 machine from the inexpensive DVD-R drives in the Apple computers: the media type used. The DVD authoring media used for recording in the high-end machine was designed for creating replication masters. The general use media used by the Apple's computers was targeted for playback on most DVD players, but not for master creation. Since then, many replicators will accept DVD general-use media for mastering (sometimes with an extra fee). Others still insist on submissions on Digital Linear Tape (DLT) or DVD Authoring media.

Apple computers in general still tend to be more expensive than comparable Windows-based machines, but, as part of a long-standing tradition, the development community still favors Apple for ease of use and trouble-free operation. Both Mac and Windows platforms these days are far more robust for optical disc recording than in the recent past, and either platform can serve you well for developing DVDs. If you've worked with computers for some time, you probably already favor either Mac or Windows systems. I'd recommend staying with the platform with which you're most comfortable.

A Macintosh system equipped for DVD recording can be purchased for around $1300. For example, a well-equipped eMac in that price range includes these features:

- eMac with 1GHz PowerPC G4 processor and 512MB system RAM
- 80GB Ultra ATA hard disk drive
- Superdrive with CD-RW and DVD-R support

All Apple computers of recent vintage include two or more FireWire ports (Apple's name for the IEEE-1394 I/O standard), so you can immediately connect any DV camcorder that includes an IEEE-1394 interface (most DV camcorders do).

Superdrive-equipped Macs also include tools for video editing (iMovie) and DVD authoring (iDVD), which can be used to create simple DVD projects. If you're going to perform more sophisticated work, you can upgrade to Apple's acclaimed NLE and DVD authoring duo: Final Cut Pro (about $995) and DVD Studio Pro 2.0 (about $500). Apple's Final Cut Express (about $300) also contains many professional features and is a good choice if you don't need all the bells and whistles of Final Cut Pro. The new version of DVD Studio Pro includes features incorporated from Spruce Technology products (since Apple acquired Spruce) and represents a very comprehensive, professionally oriented authoring product.

The range of editing and authoring applications for the Macintosh is more limited than what you can find for the Windows platform, but the available tools are very good, so you may never feel restricted. In one disappointing note, Adobe has announced that their flagship video editing tool, Premiere, will no longer be produced in Macintosh versions and their recently released authoring software, Encore, is Windows only. Other software producers, such as Avid, however, continue to recognize the importance of the Macintosh to artists and developers. The Avid Xpress Pro package includes both native Windows and Macintosh versions of the software, giving the developer maximum flexibility to freely move between development platforms

Hard Disk Drives

A capable hard disk drive is one of the more important components in a well-tuned DVD production platform. Video editing requires enormous volumes of data to be handled very quickly, both while importing content from camcorders and when performing edits of material. Digital video is captured to videotape at the rate of around 3.5MB per second. One minute of standard DV can consume 200MB of hard disk storage. The bandwidth and capacity of a hard disk drive must be sufficient to handle these data volumes and transfer rates.

When selecting hard disk drives for use in a system that will be processing video and burning DVDs, consider these options:

- **IEEE-1394 interface**: Known in proprietary terms as i.Link (Sony) and FireWire (Apple), IEEE-1394 serves as the interface standard for a number of video camcorders, as well as storage devices. Portable, external IEEE-1394 drives can be used to transport large files, such as multiple-layer DVD disc images to a replication servers. Transfer rates on IEEE-1394 drives (400 Megabits per second on first-generation drives; 800 Megabits per second on second-generation) handle imported digital video without difficulty. Support for up to 62 individual devices is provided by an IEEE-1394 computer interface.

- **Small Computer Systems Interface**: Before the advent of IEEE-1394, the Small Computer Systems Interface (SCSI) was the interface standard of choice for many professional applications, such as network storage and backup and optical disc recording. SCSI still remains popular in updated versions, such as Ultra Wide SCSI, and provides support for up to 16 individual devices on a bus. Some devices that can be very useful in DVD production, such as Digital Linear Tape drives, are typically only available in SCSI interface models.
- **IDE/ATA Drives**: The Integrated Drive Electronics/Advanced Technology standard provides support for up to four devices per computer, having evolved over several generations as an inexpensive, but capable interface standard. Low-cost IDE drives can provide efficient storage for digital video content with new models supporting very large capacities, ranging from 40GB to over 100GB per drive. IDE/ATA is not as capable of handling multi-tasking operations as FireWire or SCSI, so potential difficulties exist if several high-speed devices are combined under heavy loads on the same bus (such as a DVD burner and a primary hard disk drive).
- **Redundant Array of Independent Disks**: Out of the necessity for handling larger volumes of data common to network applicatons, a new technology, Redundant Array of Independent Disks (RAID) was developed. RAID technology incorporates a number of drives into a single cohesive unit that is available to the system as what appears to be one large drive. The multiple drive approach provides protection against individual drive failures, as well as offering a high-performance storage method that adapts well to handling uncompressed video content. Professional high-definition video camcorders often use Serial Device Interface (SDI) for content transfer, which works well with a RAID system.

Advanced Computer Options

If you decide to pursue a professional path in making DVDs, the range of equipment add-ons and accessories stretches from here to infinity. Certain kinds of options, such as Digital Linear Tape drives, may be a requisite add-on if your replicator prefers DLT submissions for mastering discs. Other types of add-ons become necessary only as you find the need to become more efficient during production and reduce the time required for common tasks. If you do a good deal of video editing, get a system with the fastest processor you can afford. Then, add a video board with a graphics accelerator and dual monitor support.

Industrial-strength storage devices such as disk arrays, increased memory, dual-processor systems, and similar kinds of add-ons can streamline many processes and reduce production time as you begin doing serious DVD production work. For example, if you intend to do your own MPEG video compression on a regular basis, an enterprise-caliber system equipped with dual Intel Xeon processors or a Power Mac G5 with dual processors can substantially reduce the time required for compression.

For up-to-date descriptions of useful products that can enhance a DV production studio, the Web offers a number of top-notch resources that include reviews of new products as soon as they reach the market. Resources that provide timely reviews include:

- eMedia (*www.emedialive.com*)
- Digital Media Net (*www.digitalmedianet.com*)
- DV Magazine (*www.dv.com*)

Selecting a Source of Content

One common misconception about the Digital Versatile Disc is that you have to have a DV camcorder to create the video content for the disc. This perceived restriction has probably stopped many potential creators in their tracks. Although consumer quality DV camcorders from Sony, Canon, JVC, and others can now be purchased for as little as $600, most moviemakers leaning towards more serious endeavors look a few steps up the feature ladder to the prosumer cameras. These generally cost between $2000 and $5000. A greater range of professional features can be found on prosumer cameras, offering moviemakers the ability to use interchangeable lenses (on some models), access a greater ranger of external devices with I/O connections, control more aspects of the image capture (through white balance adjustments, exposure controls, focusing mechanics, and so on), and support a greater range of formats, such as 24p frame mode and high-definition digital video. If you aspire to higher quality production values, aim for a prosumer or professional model. If your budget doesn't allow expense, you can still produce some very respectable video and sound with consumer-level DV camcorders.

Putting content of a DVD, however, is far simpler than you might think. Virtually anything that can be expressed in digital form—audio, images, video—can be stored and accessed from a DVD-ROM. The only restriction that you encounter when considering content for a DVD-ROM presentation is the inherent file size limitations, which should only become a factor if you plan to put video files in the multi-Gigabyte weight class onto a disc. DVD-ROMs can be thought of as giant bit buckets for your projects. You can choose your favorite video compression format—whether QuickTime or MPEG-4 or Windows Media 9. You can design autostart sequences for use with a Macintosh or under Windows. You can store complete applications, games, image libraries, MP3 audio collections, audio books, interactive multimedia works in Macromedia Flash MX or other formats, technical references, complete encyclopedias, training regimens, or just about anything else that adapts to a computer framework. Your primary consideration for this use of the medium is that you're creating works for playback on computer DVD-ROM drives; you can't insert a DVD-ROM in a set-top DVD player and start playing QuickTime videos.

You can, however, create a ROM folder on a disc designed as a DVD-Video. This ROM folder can contain all of the content available to your computer DVD-ROM drive, but these files are ignored by a set-top DVD player. As long as the file structures for the DVD-V portion of the disc are correctly placed (refer to *File Formats under DVD-Video* on page 249), the DVD player can handle the menus and submenus, as well as the video content included in the VIDEO_TS folder. You've essentially created a disc with components that permit playback through computers, as well as conventional DVD players.

If you are authoring a DVD explicitly for use in a set-top player, your audio and video files must be expressed in the language of DVD. The video must be encoded in MPEG-2 format and the audio must be encoded in MPEG-2 Layer 2 audio format or Dolby Digital (AC3) format. Other formats, such as Windows WAVE files and PCM (the native audio format of CDs) can often be converted on the fly by authoring applications, allowing their selection when compiling audio tracks for a disc.

Because camcorders do not capture video in native MPEG-2 format, there is almost always going to be some kind of conversion process involved in going from content editing to authoring for DVD. This applies whether you're bringing in video from an analog Sony Betacam camera, a Canon XL1S DV camera, a high-definition video file from a Panasonic AG-DVC80 camcorder, or a file created by a telecine suite from 35mm film. In the end, all of these sources of content get converted to MPEG-2 for DVD storage.

The graphics files that you intend to use for menus and submenus can also be acquired from a variety of sources. Some applications, such as Apple DVD Studio Pro, accept layered files, such as those produced by

Adobe Photoshop. Other DVD authoring applications may require that you flatten the files before working with them, which combines the multiple layers into a single layer.

As an example of the graphics files supported, Sony DVD Architect v1.0 lists these formats as compatible graphics media for a DVD project:

GIF: an 8-bit, 256 color lossless graphics format popularized on the Internet; uses the extension.gif.

JPEG: a lossy, true-color image format; uses the extension.jpg or .jpeg.

TARGA: a lossless image format with true-color representation that also supports alpha channel transparency; uses the extension .tga.

TIFF: the tagged image file format is supported in many different graphics applications; uses the extension .tif or .tiff.

Windows Bitmap: a popular graphics format that originated on the Windows platform; uses the extension .bmp.

Other DVD authoring applications support different graphics formats for menu building; through conversion or exporting a graphic to one of the supported formats, you can typically work with a wide variety of source materials and save to a supported format as the final step.

Selecting a Digital Video Camcorder

At the consumer and prosumer level, digital video camcorders come in two varieties:

MiniDV: Released as a consumer-oriented version of the DV format, these popular cameras have a wide following and most include IEEE-1394 interfaces for direct transfer of digital video content for editing. The small tape cassettes maintain quality through a sophisticated error correction scheme that eliminates drop-outs and flaws in recorded content.

Digital8: Sony introduced this alternative digital video format, which uses inexpensive 8mm or Hi-8 videotapes. It can also play back videotapes in these earlier analog formats, adding some versatility if you happen to have a collection of 8mm videos. Because of the size of the videocassette itself, these cameras are slightly larger than miniDV models. Many include an IEEE-1394 interface to directly input video content.

A number of more professional formats have emerged that still use the same DV signal specifications that apply to miniDV, but include extra features, such as audio that is locked to the video stream and larger media formats for improved recording quality. Panasonic introduced the DVCPRO format for news-gathering organizations and their cameras typically range from about $3500 to $10,000. Sony has advanced their popular Betacam format to encompass digital video material and named the new format Digital Betacam, sometimes called DigiBeta. Camcorder prices are roughly equivalent with Panasonic professional models. A new generation of high-definition digital video camcorders are beginning to appear with support for HD formats being added to NLE applications such as Apple Final Cut Pro and Sony Vegas v4.0

The following table summarizes the primary DV formats.

Table 1: Digital Video Formats

Property	DVC	DVCAM	DVCPRO	Digital Betacam	HD Video
Manufacturer	Sony, Panasonic, JVC, Sharp	Sony	Panasonic, Hitachi, Philips, Ikegami	Sony	Sony, Panasonic, Ikegami
Resolution	720x480 NTSC	720x480 NTSC	720x480 NTSC	720x480 NTSC	1080 lines
Color sampling	4:1:1	4:1:1	4:1:1	4:2:2	4:2:2
Audio	Unlocked 2ch 16-bit 48KHz; 4ch 12-bit 32KHz	Locked 2ch 16-bit 48KHz; 4ch 12-bit 32KHz	Locked 2ch 16-bit 48KHz; 4ch 12-bit 32KHz	Locked 2ch 16-bit 48KHz; 4ch 12-bit 32KHz	Locked 4ch 16-bit 48KHz
Tape capacity	miniDV: 120m; std DV: 180m	miniDV: 40m; std DVCAM: 180m	miniDVCPRO: 63m; std DVCPRO: 120m	mini-DigiBeta tape: 40m; std: 180m	mini-HDCAM tape: 40m; std: 180m

Features to Consider in a Camcorder

If your budget is limited to $1000 or less, you are pretty much restricted to the consumer models of DV camcorders. Sony, Canon, Sharp, JVC, and others produce a number of very competitive models in the under-$1000 category. Between $1000 and about $2500, the features and quality of the consumer-oriented camcorders become more sophisticated, with better lenses, more control over recording and exposures, a wider range of inputs and outputs for manipulating sound and video, and better construction. From about $2500 to $3500, several prosumer models from Sony, Canon, Panasonic, and Ikegami include numerous professional features, such as built-in neutral density filters and 16:9 recording options, as well as more precise image and sound control. Upwards from $3500, you can now purchase fully professional models in several categories, including DVCAM, DVCPRO, Digital Betacam, and high-definition video. These higher end camcorders are suitable for broadcast work, industrial videos, documentaries, and even feature-length moviemaking.

Some of the factors to consider in the purchase of a digital video camcorder include:

CCD: The size and number of charge-coupled devices (CCDs) in a camcorder bears a close relationship to the overall quality of the captured image. These very compact electronic sensors interpret the intensity of light (luma) and the color (chroma) according to a sampling grid and turn this information into digital values. The consumer-level camcorders typically use a single CCD measuring 1/6". At the prosumer level, three CCDs provide more accurate representation of the color spectrum—these CCDs usually measure 1/3". Professional camcorders use three CCDs that measure 1/2" to 3/4", clearly boosting the precision of the image capture. The optical sensor resolution is another value that indicates the relative image sensitivity of the CCD. For a typical prosumer camera, this value is typically about 380,000 pixels.

I/O Connections: Being able to easily move digital video and audio content in and out of a DV camcorder is essential to high quality editing. The more options that you have for inputs and outputs, the more flexibility you will have in your video work. The well-established standard for high-speed video transfer is IEEE-1394, which is called FireWire by Apple and i.Link by Sony. A single cable can link your digital video camcorder to your IEEE-1394-equipped computer and not only transfer the video, but control the operation of the camcorder from your computer (providing fast-forward, reverse, pause, and so on). Panasonic did not offer this interface initially with their DVCPRO camcorders, much to the distress of editors and video professionals. Another I/O capability that can prove useful is support for a three-pin balanced microphone input, allowing professional microphones to be cabled to the camcorder. On some prosumer camcorders, such as the Canon XL1S, you can purchase an adapter that provides balanced mic inputs. Other connection inputs and outputs can be handy: S-Video, USB connections for reading Memory Stick Digital Media values, and so on. Think about the types of equipment with which you want to interface. Then, check the specifications for your selected camera to see if it supports the required devices.

Still Image Capture: Many camcorders have a mode where they can be used to capture still images, although the captured images are typically of lower resolution for use primarily with email or Web postings. Stills can generally be captured at 640 x 480 or 720 x 468, if this feature is available. If being able to capture both still images and video is a significant consideration for you, check this feature availability on your selected camera.

Interchangeable lenses: One of the primary separators between the world of consumer camcorders and professional camcorders is the ability to change lenses when required. Particularly in feature film work, the base or stock lens for a camera rarely has the versatility to handle the range of visual effects demanded by the director's approach. Cameras such as the Canon XL1 brought a new level of professionalism to the DV world with support for interchangeable lenses (for an example of this approach, see Chapter 15, *Case Study: Creating a High-Caliber DV Short Film*). Some camcorders can be equipped with an adapter that allows use of high-quality lenses from 35mm photography; this capability greatly enhances the utility of the camcorder, but realize that lenses and accessories can sometimes cost as much or more than the camcorder itself.

Optical and Digital Zoom: Most camcorders equipped with a zoom lens provide a value for the optical zoom range and a value for the digital zoom range. The optical zoom range (for example, 12x for a Sony DCR-VX2000) indicates the lens to focal length adjustment for which the camcorder is capable; the image quality throughout the zoom is equivalent to the maximum optical quality of the lens. A digital zoom value (48x for the same Sony camcorder) indicates a simulated digital zoom that loses resolution as the image is expanded. For quality work, the optical zoom value is the more important value to consider.

Weight and feel: If you intend to do frequent work with a DV camcorder, the weight, design, organization of controls, and other tactile features can become very important very quickly, particularly on larger units that use a shoulder rest. This is where some hands-on testing at a photography store or a video retail establishment can be essential. Check the balance, the heft, the viewfinder, the ease of use. Does the camcorder have a manual focus ring? Servo-controlled focus rings, such as used on the Canon XL1S camcorder, can be difficult for those used to focusing 16mm or 35mm motion picture cameras.

Reputable sources for digital video camcorders include:

Computer Discount Warehouse: *www.cdw.com*

B & H Photo: *www.bhphotovideo.com*

PC Connection: *www.pcconnection.com*

Examples of digital video camcorders include:

Sony DCR-VX2000: This Sony DV Handycam employs three, 1/3″ progressive scan CCDs, producing a horizontal video resolution up to 530 lines. Images to be recorded can be viewed through a 2.5-inch, 200,000-pixel LCD screen that swivels or a 180,000-pixel through-the-lens viewfinder. A frame-recording mode provides progressively scanned images and 16:9 widescreen recording is also an option. Sony's effective optical picture stabilization is built into the unit for more professional results when hand-held camera movements are made. Sony's i.Link interface (equivalent to IEEE-1394) is included for seamless DV transfers to a computer. The typical retail price for this camcorder is about $3000.

Canon XL1S: The first prosumer-class camcorder to introduce support for interchangeable lenses, Canon XL1 series camcorders have been widely adopted by a new generation of moviemakers. Three 1/3-inch CCDs provide precision color through Canon's pixel-shift technology with images delivered through very high quality optics. This Canon camcorder supports a Frame Movie mode, which records non-interlaced images at 30 frames per second. A wide variety of after-market accessories can be purchased for this camcorder, including a number of lens adapters that can significantly enhance the optical capabilities, making it a solid choice for DV moviemakers who want to experiment with lens selections. The Canon XL1S sells on the street for around $4200.

Selecting a Duplicator

One key piece of equipment that can change the way you work is an optical disc duplicator. Particularly if you are frequently performing short run duplicates of projects in progress, a duplicator can free you from dependence on large-scale replication services and the tyrannical limitations of minimum runs. Ideally, a duplicator can also serve as a revenue-generating tool if you decide to market your DVDs directly to the public. DVD duplicators range in price and capabilities, starting from simple standalone models that can produce three or four discs an hour to automated units with full robotics and integrated label printers that can work unattended on 50 to 100 DVDs per session.

Before you can make a decision on what duplicator model works best for you, you need to honestly assess the volumes of DVDs that you will be producing. If you only need to generate ten or twenty DVDs a month, a large-scale automated disc duplicator is generally not necessary. Standalone duplicators that function with the simplicity of a photocopy machine can probably serve your needs. If you find that you're needing twenty, thirty, or more discs every month, one of the low-cost automated duplicators could best serve your needs. If your production needs reach as high as hundreds of discs a month, a high-end duplicator with multiple drive bays will reduce the amount of time you spend on duplication tasks. Most of the automated duplicators permit the attachment of a printer, which can be integrated into the robotics system so both the duplication and printing become fully automated.

An added bonus of the current generation of duplicators is that most of them will also burn the full range of CD formats, so you can use them for double duty, generating both CDs and DVDs from the same system.

DVD-R or DVD+R?

Many duplicator manufacturers and distributors offer the option to select the type of DVD recorder to include in a duplicator system. The two types available are:

- **DVD-R**: The original write-once format sanctioned by the DVD Forum, DVD-R comes in two variations according to the type of media it uses: authoring use and general use. Because this format has been in existence longer, compatibility with earlier generation DVD players is better. For more details, refer to *DVD-R* on page 255.
- **DVD+R**: A relatively new write-once format developed by the DVD+RW Alliance. This format was introduced to improve player compatibility and slightly boost recording speeds. Most of the latest generation of DVD players read DVD+R media, and many older players as well can also handle the media.

In practical applications, both DVD-R and DVD+R have extensive compatibility with most modern DVD players, typically in excess of 94 percent for both versions with DVD-R having a slight edge. Earlier compatibility for both formats is more problematic. The tradeoffs between these formats make them relatively equal for most purposes. When choosing media, you need to match the type of media to the recorder type: DVD-R media for DVD-R recorders and DVD+R for DVD+R recorders.

Features to Consider in a Duplicator

The following features and options apply to current generation DVD duplicators.

- **Manual or automated**: Manual DVD duplicators, though less expensive, require an operator to physically insert discs into the drive and remove them following copying. An automated duplicator can burn discs from one or more masters and a spindle stacked with recordable media. If your duplication needs exceed twenty or thirty discs a week, you might want to consider an automated unit.
- **Recording speed**: A 4x DVD-R drive can record about four DVD-R discs per hour, assuming the master disc is close to full file capacity (4.7GB). Earlier 2x DVD-R drives can only achieve half this output. If you record less than full capacity, however, duplication takes less time. DVD+R drives often have a 2.4x recording speed.
- **Integrated printer**: Many of the automated DVD duplicators include integrated operation with an inkjet or thermal printer, designed to apply the labeling after the disc has been copied. With an integrated printer, you can automatically copy and label discs, a necessity in any kind of high-volume production environment where a professional appearance is important. Inkjet printers can handle four-color surface printing that includes elaborate graphics or photographs. Thermal printing is less expensive, but is more suited to printing text or simple logos in black.
- **Dedicated PC**: The processor-intensive nature of DVD duplication generally requires a dedicated PC to oversee the operation. Some duplicators require that you dedicate a PC to this process, which means that you must forego any other operations on this PC while duplication is taking place. Other duplicators include a built-in processor and can operate independently from any other system. If you are purchasing a duplicator that requires its own dedicated PC, factor this cost into the overall cost of the unit when comparing it to systems that have built-in processing.

- **Upgrade potential**: Some duplicator systems support an easy upgrade path. For example, you might initially purchase the duplicator unit and then later add an integrated printer that can be accessed with robotics. Or, the initial duplicator might feature six drive bays with only two of these bays populated with DVD recorders. When you need to upgrade your production rates, you can add two or more drives to the main unit. Some units also have upgradable firmware, allowing you to obtain patches or new firmware revisions to accommodate future changes. These types of units are generally more flexible and long-lived than units that don't support firmware upgrades.
- **CD support**: Although most DVD recorders support recording to CD media, as well, the recording speeds are typically much slower than a CD-only duplicator. A typical 4x DVD recorder can record CDs at 16x speeds, as compared to 40x or 48x for a CD-only unit. If high-speed CD recording is a necessary element of your production environment, you might want to consider purchasing a separate CD duplicator to obtain the faster recording speeds.

Examples of DVD duplicators include:

- **Disc Makers Reflex1**: This unit can be equipped with either a DVD-R drive (for 4x recording speeds) or a DVD+R drive (for 2.4x recording speeds). The simple operation is similar to a copy machine—you insert a master in the upper drive and blank media in the lower drive and press the start button to make a copy. As a low-priced entry to the world of disc duplication ($890), this unit offers a good deal of value for the money. For details, visit: *www.discmakers.com*.
- **Alera Technologies 1:7 DVD Copy Tower Dual DVD/CD Duplicator**: Simultaneously duplicates up to 7 copies of a DVD disc at speeds up to 4x. Designed as a standalone tower unit, this duplicator handles multiple formats, including DVD-R, DVD=R, DVD-RW, and DVD+RW. List price at price time was $3,099. For details, visit: *www.aleratec.com*.
- **Disc Makers Elite2**: The Disc Makers Elite2 integrates two 4x Pioneer DVD-R drives and an inkjet printer to produce an output of 8 DVDs per hour (color printing included). Mechanical robotics can handle unattended operation for up to 125 discs at a time. A dedicated PC is required to oversee operations. The current retail price for this duplicator is $5590. For details, visit: *www.discmakers.com*.

Selecting a DVD Recorder

You can make very respectable DVDs without owning a DVD recorder. Many of the DVD authoring packages provide an emulation feature, so you can play back video from the menu being designed without having to burn a disc. Once the structure and organization of the DVD contents have been defined, the files can be transferred to DLT using the data description protocol (DDP) for submission to a replication service.

Realistically, however, if you're making DVDs as a part of your business or as an avocation, having a DVD recorder will add immeasurably to your work. You can store interim copies of movies that are being edited. You can produce test copies of DVD projects for quality control reviews and client approval. You can create a library of key assets for different projects for near-line storage. Given the modest expense of a DVD recorder (ranging from $250 to $600 for many general-use DVD-R and DVD+R units), this single piece of equipment can make a very big difference in your workflow.

Authoring Use DVD Recorder

The granddaddy of DVD-R authoring, the **Pioneer SVR-S201**, was the first DVD-R recorder to handle 4.7GB media, as specified in the DVD-R Book Version 2.0 for Authoring. This drive was initially introduced (in July of 2000) to streamline the process of submitting DVD mastering content to replication services, making it possible to deliver files on disc rather than on tape. Support for the Cutting Master Format (CMF) also helped in eliminating potential problems in working with replicators and to refine a process by which a final authored DVD-R (produced on authoring media) could be tested and then directly submitted to the replication service. A set of data values are recorded in the read-in area of the DVD-R media to convey appropriate information for producing a glass master and pressing discs.

The SVR-201 uses a Fast SCSI interface and records at 1x speeds, producing a completed DVD-R in about an hour, including the time required to verify the files written to the media.

This SVR-S201 still enjoys significant support in the DVD authoring community, despite its steep price tag (with a street price of about $4000 through retailers such as B & H Photo) and relatively slow recording speed.

A number of replication services now accept content on General Use DVD-R media, but additional work is required to process the master files, handle the addition of copy protection (if required), and verify the DVD disc image. Though this process has been automated to some degree, it creates extra work for the developer and the replication service and introduces a degree of uncertainty into the replication process.

External DVD Recorder

If your computer is not already equipped with a DVD recorder, it is a relatively simple process to add an external unit. If you have an additional drive bay and room on your ATAPI/IDE chain, you can also add an internal unit, but this requires a bit more work to configure. Many external DVD recorders offer the flexibility of dual interfaces, providing both IEEE-1394/FireWire and Hi-Speed USB options. This simplifies installation and connection, as well as offering dependable disc recording.

Examples of external DVD recorders include:

Plextor PX-504UF: This drive features both IEEE-1394 and USB v2.0 connectivity; it can be used both with Macintosh and Windows PCs. DVD+R and DVD+RW media can be used, with a 4x recording speed. CD recording at 16x speeds in a variety of formats can also be performed. The street price for this unit is about $315.

Sony DRX-500ULX: Sony has dealt with the confusion over different DVD recording standards by creating a drive that supports all of them. The DRX-500ULX handles DVD-R, DVD-RW, DVD+R, DVD+RW, as well as the full range of recordable CD formats. Designed with interfaces for IEEE-1394 and USB v2.0, this recorder adapts well to both Macintosh and Windows applications. DVD write speeds are 4x; CD recording can be performed at 24x. The unit typically sells for around $375.

Standalone DVD Recorder

A new class of DVD recorders has emerged, designed to sit on top of the television and capture programs to DVD. Processing advances have made it possible to encode MPEG video files on the fly and simultaneously record to DVD. The OpenDVD standard has made it possible to work with the content being recorded to DVD in a progressive manner, adding new material and creating a structure that makes the video files accessible for editing or transfer. Sonic Solutions devised this approach and developed a scripting language, AuthorScript, that supports the open framework. The OpenDVD specification supports both recordable and rewritable DVD media.

Besides simply recording television programs, these recorders can be useful in a production environment because most contain high quality hardware encoders that can accept video streams from externals sources and convert it to DVD formats in real-time. This is essentially the same function that is provided by a video capture board that is installed in a computer, but in this case the video files are being recorded to DVD rather than being copied to a computer's hard disk. One possible disadvantage to the encoding process is that the bit rates are typically fixed at less than the optimal bit rates that are possible for a given segment of video. This can produce slightly lower quality video content than if the encoding was handled by a two-pass variable bit rate encoder. Overall, however, video encoded with these recorders is very satisfactory for viewing and standard applications.

If you are regularly relying on input from analog sources to create DVD projects, such as video from Betacam SP video cameras, a standalone DVD recorder can be a valuable asset in your production studio.

Examples of this type of unit include:

Sony RDR-GX7: With support for DVD+RW, DVD-R, and DVD-RW, this Sony recorder provides significant flexibility. A built-in TV tuner simplifies the task of recording broadcast material. Features component, composite, and S-video outputs. The street price for this unit is around $700.

Philips DVDR75: This recorder supports recording to DVD+R and DVD+RW media. Playback capabilities include DVD+R, DVD-R, commercial DVD-V titles, Super VCD, Video CD, audio CD, CD-R, CD-RW, and MP3 CD. A IEEE-1394 connector makes it easy to transfer captured video to a computer. The street price is about $500.

Renting Equipment for Projects

Many DVD projects, particularly those that include live videotaping of events, documentaries with on-site shoots, short features, and full-length feature films, will require more equipment than you have available. Purchasing everything that you need for such a project can often push your budget way out of line with your expectations. In such a case, renting may offer a reasonable solution. Most metropolitan areas offer rental companies that can meet the requirements of a videographer or independent filmmaker; even some smaller communities include at least one option (see Chapter 12, *Case Study: Supporting Independent Filmmaking* for an example of a small community with extensive production support).

The availability of rental equipment can significantly boost the quality and results of a project, allowing you to incorporate top-notch microphones and sound gear, stabilizers for camcorder movement, professional lighting gear, better quality motion picture cameras or camcorders, and similar items. To be cost-effective, equipment rentals should be planned carefully, both for budgeting and scheduling (see Chapter 3 for more information).

Selecting a DLT Drive

DVD-R Authoring media can be used for submitting a DVD master to a replication service for DVD-5 format discs (single-layer), but what do you do if you're creating a higher capacity DVD, such as a DVD-9? Most replication services favor submissions on Digital Linear Tape (DLT), an output format that is supported by many of the DVD authoring applications, including Apple DVD Studio Pro v2.0. The use of DLT as a submission media allows developers to incorporate the full range of DVD features into a title, including CSS copy protection, region coding, and so on.

DLT drives first gained popularity as high-capacity backup devices for enterprise-scale networks. Individual tape cartridge capacities range for 20GB to 320GB (compressed). Prices also vary widely, ranging from about $1500 to $8000. Often, equipment being retired from network backup use or other applications can be purchased on eBay for a small fraction of its original cost. This less-expensive alternative may prove useful if you are only producing a few titles for replication per year.

If you are purchasing a DLT drive for your production studio, many of these units include only a SCSI interface (typically, Ultra Wide SCSI). As most mainstream computer systems no longer include a SCSI host adapter, you may need to add this additional hardware to be able to connect the DLT drive.

Examples of DLT drives currently available include:

Quantum DLT1: This DLT drive accommodates a typical small to medium-size workstation with a capacity of 40GB uncompressed/80GB compressed storage. Designed to provide read compatibility with the DLT 4000 format, the DLT1 enjoys wide system support in many backup applications and hardware platforms. Backup speeds in native format are 3MB per second; in compressed mode, 6MB per second. The street price of this unit is $1050.

Quantum SDLT 320: This tape drive takes advantage of SCSI LVD transfer rates to achieve a backup rate of 115GB per hour. Native tape capacity (uncompressed) is 160GB. The compressed tape capacity is 320GB. The SDLT 320 tape drive employs both magnetic and optical techniques to improve capacity and performance. The street price of this equipment is $3800

While used DLT equipment purchased on eBay or other online auctions can be a bargain, many of these auctioned items do not include documentation, software drivers, cables, or anything other than a box, which may or may not have been tested prior to being placed on auction. Shop cautiously for equipment in this venue and pay close attention to the feedback ratings for those offering the equipment.

Establishing a Production Studio

Equipment choices are highly personal selections and your needs and requirements are bound to change over time. If you buy computers, camcorders, storage devices, and other accessories slightly behind the technology curve, you can often save money. If you're working in a highly competitive professional environment, however, you may need the most modern equipment available to keep stride with others in your industry.

The technology trap can be endless (and the expenses ongoing), so to some degree it is wise to try to take maximum advantage of the tools that you have in hand, rather than waiting for the inevitable next-generation model that adds one more must-have feature. You can accomplish remarkable work with even fairly modest DV equipment and you can create DVDs to communicate and entertain without needing the resources of Sony or the technical staff of Zoetrope to produce your material. Use your tools wisely, as described in the following chapters, and you can gain exceptional results, whatever your level of equipment.

3

Planning a Project: Making Things Happen

A DVD project can range from a simple compilation of your favorite nature photographs accompanied by original guitar music to a full-length feature film with a large cast and crew and nervous investors. Either way, whether you're doing work for a client or creating your own production, planning and budgeting can make things go much smoother. The complexity of even a fairly simple DVD project can tax your ability to carry all the information around in your head and make instant decisions on budget adjustments and scheduling changes. Even if you don't consider yourself an organized person and you think that a true artist never commits a process to paper, you may be surprised how much a bit of planning can actually free you to be more creative on a project.

The film industry generally divides projects into three major categories: pre-production, production, post-production. If you're working on a DVD project that involves primarily video or feature film material, these three categories will have significance to your work as well, but you will also have a number of other DVD-related workflow steps. Some of these could be considered within the category of post-production, but to illuminate the process, we'll use terminology more closely aligned with the optical disc industry. Following post-production, the three DVD stages that we'll discuss include authoring, pre-mastering, and mastering.

On most projects you will be coordinating your work with a number of other specialists: sound engineers, audio editors, video editors, rental facilities, cast members, production crew members, replication facilities, and so on. Busy, talented professionals in these fields don't like to have their time wasted while you contend with the after effects of bad planning. Even if you're working with volunteers and using barter to obtain services, if you respect the time of those you are working with, they will be far more likely to work with you on other projects down the road.

Plans and budgets are guaranteed to change along the way. In any project, there are always unpredictable elements: the unexpected weather conditions that cause you to postpone a shoot, the lead actor who comes down with the flu on the second day of shooting, the lighting crew whose truck breaks down on the way to the sound stage. You'll be better equipped to deal with these unexpected contingencies if you've worked out alternate plans and forecast expenses in a realistic way, leaving some headroom to cover unanticipated costs and slack time to make up for slips in the schedule.

DVD Production Process

If you can visualize the entire DVD production process, from beginning to end, it is much easier to start planning the individual steps needed to launch the project. There are as many ways to look at the DVD production process as there are DVD titles, but, in general, the following stages are involved in any DVD project, as shown in Figure 3 - 1.

Figure 3 - 1 **DVD Production Steps**

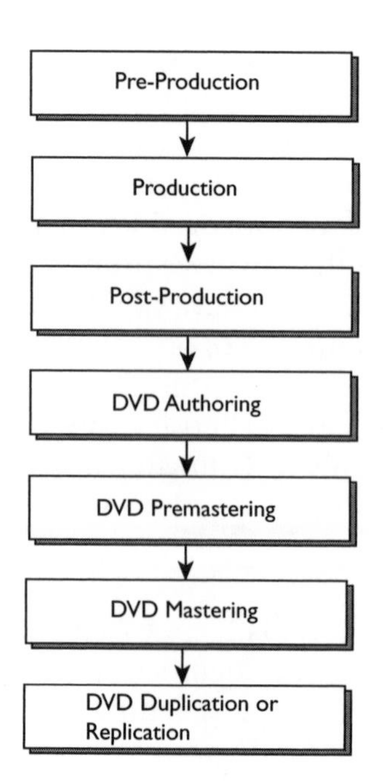

- **Pre-Production**: The pre-production stage of a DVD project represents the entire planning process, from devising an initial concept to planning the logistics. At this stage, you need to consider all of the elements needed for your particular type of project, including personnel, equipment, audio or video assets, script, storyboard, and so on. Budgeting, of course, is also a key element at this stage. Establishing a realistic budget for a DVD project will help ensure that a project can actually be completed and help you set milestones to track expenditures.
- **Production**: The production stage involves all of the work required to create the core content for your DVD. If you're doing a documentary, it includes the interviews and location shooting. If you're doing an animated short, it includes the artwork production and the cel or motion animation. If you're doing a feature film, it includes all of the work required to shoot the scenes for the movie, whether in film or video format, in preparation for editing.
- **Post-Production**: By the time you've reached the post-production stage, you should have the majority of the necessary assets to complete your project. At this point, non-linear editing tools and sound editing applications create the flow of your production, layering tracks, adding and subtracting elements, refining the storyline, enhancing the content. In a full-blown feature film, this stage could involve dozens of specialists, from audio engineers to video editors to animators. If you're doing a small-scale production, the post-production tasks may fall entirely on your shoulders. Typically, the post-production phase of any project is far longer than the production phase. In other words, if it takes you three months to shoot an independent feature film, it may take you nine or ten months to complete post-production. In many cases, you will also end up taking journeys back to production tasks: reshooting scenes, rerecording dialog in a sound studio with looped segments of film or video. At the conclusion of the post-production process, you should have a movie or a documentary or an animated work or a series of instructional videos to plug into the DVD title.
- **DVD Authoring**: DVD authoring involves all of the steps required to organize and present the assets that you created in the first three stages of the production process. If your title includes a mix of music and video (such as might appear in a monthly DVD music video magazine), you typically would be creating a menu

similar to the table of contents in a magazine. If you're producing a feature film, you would be designing chapter entry points and assembling sub-menus that offer access to the DVD extras. The authoring process is the design phase where you essentially create an interface to view the assets that will be on the disc. This process is explained in more detail in Chapter 8, *Authoring a DVD Disc: Creating a Title,* which also provides some examples of authoring applications in use.

- **DVD Pre-Mastering**: This stage of the process leads to the production of a disc image that reflects the DVD file structure and organization. The disc image includes all files converted to their DVD-ready formats. For example, the video assets are encoded, converted to MPEG-2 format and the audio assets are converted to Dolby Digital or MPEG-2 audio format. Error-correction codes are created and audio and video streams are multiplexed together. The disc image can be used as the basis of replication or duplication. The image may be stored on a portable hard drive for transport, burned to recordable DVD media, or transferred to digital linear tape for delivery to a replication facility.
- **DVD Mastering**: Mastering describes the process by which a master copy of a DVD is created, either by a manufacturer for replication or by a developer creating a DVD disc to be used for duplication. In both cases, the master serves as the final bit-for-bit version of the DVD which gets used to make stampers (in the case of manufacturing) or becomes the master disc inserted into disc duplication equipment.
- **DVD Duplication or Replication**: From a disc master, the duplication or replication process can be initiated. In the case of duplication, recordable media is used to write a copy of the master to disc. In the case of replication, the manufacturing sequence creates the specified number of pressed discs, using an injection molding process by which melted polycarbonate beads flow into a stamper where they are imprinted with the pits and lands composing the disc data.

These processes are discussed in more detail in the chapters to follow.

Developing the Concept

A highly popular urban myth credits the idea behind a multi-billion dollar corporation, Company XYZ, as originating on the back of a cocktail napkin during a dinner table conversation. In reality, this idea is less of a myth than you might imagine. A number of successful Silicon Valley computer companies were launched from chance conversations, idle dinner discussions, or casual brainstorming over cocktails. The founders of companies such as Hewlett-Packard, Apple Computer, and Microsoft Corporation started out with simple ideas and grew them into internationally recognized icons. Good ideas can have a momentum of their own, but the number of steps separating the idea from the physical manifestation of that idea can be enormous. Persistence, vision, and dedication are needed to bring any idea into existence, and creating a DVD title can require a significant investment of time and energy.

Brainstorming

The time-honored process of brainstorming plays a role in many creative endeavors. Unless you are creating a DVD title on your own, you'll probably spend some time with one or more co-creators, mapping out the direction of the project, choosing the techniques you will use during development, discussing your potential audience and possible distribution channels.

Some simple ideas can make brainstorming more effective:

- Find some means of capturing the creative flow as it is happening. You can make an audio or video recording of the discussion as it is taking place (video can be handy if the brainstorming includes diagrams drawn on a conference room white board). You can have someone assume the role of secretary and jot down the ideas that are being expressed. The key idea is to keep the ideas flowing and be able to return to the best points later. You'll be surprised how much you can forget even one day after a brainstorming session. The best idea might have been the one that slipped out of your memory.
- Don't criticize anyone's ideas while brainstorming. Silencing the critical faculty is a primary means of encouraging the mind to be open and explore new approaches. Often times a thought or comment that seems completely crazy or totally off track will contain the seed of an idea that will brilliantly solve a problem or lead to a totally original approach. Nothing damps a spirited discussion

more quickly than someone cutting in while the ideas are flowing giving you all the reasons why the idea can't possibly work.

- Act on the results of the brainstorming quickly. A tremendous amount of energy and enthusiasm can be generated during a brainstorming session with intelligent, creative people. Take advantage of that energy and choose a next step, some course of action that will move the idea further along. Whether it is sketching out a storyboard, writing a treatment, producing a rough script, or developing a simple prototype of a DVD menu, the more rapidly the concept begins to take shape, the easier it will be to maintain the momentum.

If your concept and project require additional support, either in terms of funding, a corporate blessing, or committee approval, your first action item should be devising the strongest means of presenting the concept to a group or individual. Many different presentation tools exist and the choice of tools isn't as important as making sure that your concept resonates clearly and targets the requirements of your audience. Software products such as Microsoft PowerPoint, Macromedia Flash MX, or any competent HTML editor can help you put together a respectable presentation to win over your audience.

If you're aiming for the public sector with a DVD project that might be eligible for one or more grants, there is typically a formal structure that is favored by most foundations providing money for projects. Learn the requirements of the foundation, paying particular attention to the types of projects that they support, before investing too much energy into crafting a grant proposal. A number of excellent references on grant writing exist, both on the Web and in bookstores and libraries. One good bet is:

- **I'll Grant You That**: This step-by-step guide to grant proposal writing and funding options includes a CD-ROM with sample proposals and useful resources for those trying to identify the best potential funding sources. Written by Jim Burke and Carol Ann Prater, this book was published by Heinemann in 2000. More details can be found at *www.heinemann.com*.

Scriptwriting and Storyboards

Once the initial project concept has been established and team members are eager to start work, a script and/or a storyboard can guide the process forward. Whether you write the script on an old Olympia manual typewriter or a Macintosh iBook running *ScriptWare*, the fundamental precepts hold true. Similarly, storyboards used in the motion picture industry to create a sequence of scenes have used nearly every media imaginable, from rough pencil sketches on newsprint to elaborate, full-color acrylic paintings. A storyboard can convey scenes unfolding with stick figures that illustrate camera angles or intricate sketches that show the expressions on each character's face. Don't be concerned about producing storyboards that are works of art, as much as a convenient, effective means for coordinating the efforts of a director, actors, and film or video crew. Software tools are also available for storyboarding, some standalone, others embedded in other software. Adobe Premiere includes a built-in story editor that can use either sketches or captured stills from a video sequence to present and advance the storyline.

If you're writing to express the ideas for your own production, you may not be as interested in adopting software that meets the latest Hollywood formatting standards. However, if you will be working with experienced professionals who are used to the standards developed over the last few decades in Hollywood, a software product that produces familiar, formatted scripts can help ensure that seasoned staff members will be comfortably familiar with the presentation of the script.

Some of the popular scriptwriting tools available include:

- **Script Wizard**: This add-on template for Microsoft Word supports 12 individual script formats, ranging from TV sitcoms to feature films. Those familiar with Microsoft Word may prefer an approach that lets them leverage their experience with a word-processing program, rather than having to learn an entire scriptwriting application for scratch. The retail price is $149.95. More details about Script Wizard can be found at *www.warrenassoc.com*.
- **Scriptware**: This standalone scriptwriting product includes both Mac and PC versions and flexible options for script types. Designed to be highly automated, it formats script elements as it goes, using the Tab key to jump from common functions, such as scene descriptions to character dialogue. With a number of import functions to allow scripts from other similar programs to be accessed and modified, Scriptware is a capable, useful tool for many different types of scripts. The retail price is $199.95 (with a $99.95 competitive upgrade option). More details about Scriptware can be found at *www.scriptware.com*.

- **Movie Magic Screenwriter 2000**: This highly acclaimed software tool simplifies the writing process while streamlining format operations. It has won favor with writers from Francis Ford Coppola and Wes Craven for its intuitive approach to the process and flexibility. A number of output formats are supported, including an option to generate Adobe Acrobat files, which can be helpful for distributing scripts to cast or crew who may not have the original application. The retail price is $229.95. More details on Movie Magic Screenwriter 2000 can be found at *www.screenplay.com.*

- **Final Draft 6**: Another of the stalwart tools adopted by many successful mainstream screenwriters, Final Draft 6 produces scripts for television or film, with additional formats that support stage productions. With Mac and PC versions available, this software uses an interface that is close enough to Microsoft Word that the transition to a standalone application should not be too difficult for Word-savvy writers. Includes a number of built-in collaboration tools that could be helpful in getting a DVD project launched and executed. The retail price is $199.95More details on Final Draft 6 can be found at *www.finaldraft.com*.

For storyboarding, here are some of the available options:

- **StoryBoard Quick v.4**: This storyboard tool includes a set of predrawn characters that can be arranged in a variety of action poses to map out the essential elements of a storyline. Designed as an inexpensive, easily mastered approach to storyboarding, you can create scenes using a variety of aspect ratio presets that correspond to wide screen formats, high-definition TV, video, and European film. This retail price is $279.99. Additional information about this product can be viewed at *www.powerproduction.com*.

- **StoryBoard Artist**: A more professionally oriented storyboard tool, StoryBoard Artist functions as a multimedia design tool that can handle everything from interactive presentations and training to animatics of films. Synchronized sound can be added to presentations and transitions can be used to provide more elegant sequencing of a procession of scenes. One mode allows created content to be easily uploaded to the Web for viewing and feedback. The retail price is $799. A complete description of Storyboard Artist can be accessed at *www.powerproduction.com*.

- **BoardMaster Storyboard and Timing Software**: Provides an inexpensive, flexible means of developing storyboards using imported images or created illustrations. Zooming, panning, automation, and other cinematic features are supported, making it easy to visualize the final presentation. An export feature converts the storyboards to HTML format for review or feedback. The retail price is $99.95. Additional information is available at *www.boardmaster-software.com*.

Planning and Budgeting Tools

Enterprising software developers rarely leave any niche uncovered when it comes to designing applications. A fair number of specialized project planning and budgeting tools exist that are targeted at producers of documentaries, film, or video projects. These tools can be expensive (sometimes as much as $1000 for a single-user license), but they can also be helpful if you are new to the exigencies of real-world production work. A sampling of the applications in this realm includes:

- **Movie Magic Budgeting**: Widely adopted by film studios and professionals, this application includes both Mac and Windows versions. The features tend to favor larger productions and this product may be overkill if you are generally working on smaller-scale projects. Cost is approximately $700. More details can be found at the Entertainment Partners Web site: *www.ep-services.com*.
- **Cinergy 2000 MPPS**: The Cinergy 2000 Motion Picture Production System uses a module-based organization, with individual modules that cover scheduling, budgeting, on-set shot logging, labor rates, video assist, and post production. This approach lets each individual company supplement the application as needed to handle a wide range of production types, such as large-scale feature films, television episodes, or weekly sitcoms. The main application costs $399 and individual modules vary in price. The product runs under Microsoft Windows. More details are available at the Mindstar Products Web site: *www.mindstarprods.com*.
- **EP Budgeting**: From the folks who brought you Movie Magic Budgeting, EP Budgeting is a simplified version of their flagship product. Nonetheless, EP Budgeting includes Mac and Windows versions, compatibility with Movie Magic Budgeting files, the ability to manage multiple budgets within a single production, and a viewer utility that lets users without the program examine file con-

tents. The product costs $699. More information can be found at: *www.entertainmentpartners.com*.

Advancing the Concept

Brainstorming, scriptwriting tools, and storyboards can help refine your production goals and project techniques, offering a collaborative means of building ideas and effectively advancing the production of your DVD title. As a part of the planning process, you should also be thorough about establishing a working budget and a schedule that allows sufficient flexibility to adapt to the kinds of last-minute changes that will surely arise as you begin working on your production. Lack of funds and scheduling overruns have doomed many otherwise promising projects. Do your preproduction work to exacting standards and the latter phases of your project should go much more smoothly.

4

Capturing Audio and Video: Shooting Successfully on Location

Shooting on location often requires adapting to conditions that tax even experienced production crews and veteran DVD producers. Depending on the nature of the project, a location shoot might involve indoor or outdoor camera work, as well as audio recording in the field under varying conditions, ranging anywhere from crowded, noisy city streets to a windy, rain-soaked coastal beach. Whenever outside the studio environment, the quality of the video and audio content depends on attention to a number of factors. Ambient sound, recorded dialogue, interior and exterior lighting, the nature of the surroundings, and other factors come into play when trying to capture the highest quality recorded content.

The video content on DVD undergoes a lossy compression process, which results in a small but detectable degradation of the image quality. To ensure the highest quality video content for DVD use, you need to work diligently throughout the production process to obtain optimal quality recorded video. While there is much that can be done, particularly using computer techniques, during the post-production process to improve video content in a number of ways, you can't effectively correct badly overexposed or underexposed images where detail is lost. In this respect, video is much less forgiving than film, where adjustments during the development of the film can compensate for exposure problems.

Many modern digital video camcorders work exceptionally well in low-light conditions, as well as natural light conditions. At the same time, your production can gain a significant boost in overall quality by applying classical cinematography techniques during videotaping to light the scenes according to the effects and aesthetics that you want to achieve. As described in Chapter 15, *Case Study: Creating a High-Caliber DV Short Film*, the results can often be very impressive when professionals apply their expertise to image capture, sound work, and staging. This chapter discusses the approaches that you can use when capturing audio and video on location to achieve optimal results.

Controlling Camcorder Movement

The availability of small lightweight video camcorders has helped change the aesthetics of moviemaking and video production. Television series, such as Barry Levinson's *Homicide: Life on the Streets*, took advantage of handheld video cameras to produce an edgy, fluid sense of movement, in contrast to the precise controlled camera movements customary with heavy equipment mounted on dollies and cranes. Handheld movement can be very effective with certain kinds of documentaries and even some categories of feature films where a harder portrayal of reality is desired, but for many kinds of work, you will want more exacting control over the camera movement.

The easiest and least expensive way to control camcorder movement is by using a tripod. This simple piece of equipment can enable you to do smooth pans, executive precise vertical movements, and, with some models, perform camera movements that encompass a full range of motion.

More professional tripods use modular components and you can substitute different accessories to accomplish different ranges of movement. Tripod heads, in particular, come in three basic varieties:

Pan and tilt: Usually associated with the less expensive tripods, pan and tilt tripod heads generally feature two separate levers that control horizontal and vertical camera movements. These units typically use a friction mechanism to control the amount of force required to move the head across the selected plane, which must be carefully adjusted to obtain smooth movements.

Ball head: A tension knob adjusts the pressure on a ball affixed to the camera mount, allowing a full range of motion to be incorporated. This range of movement allows considerable flexibility when videotaping, but it is more difficult to control the camera movement along a single plane, such as when you want to perform a pan.

Fluid head: The most professional motion can be obtained through the use of a fluid head, which contains either a gas or a fluid to regulate the motion of the camera. As might be expected, a fluid head costs more than a ball head or pan-and-tilt head, but the improvement and consistency it adds to videotaped movements can add a measure of polish to your production.

Heavier tripods with a wide range of adjustment points and a wide base can provide additional stability, particularly when working with prosumer and professional camcorder gear. Don't get into trouble by trying to put a 35mm still camera tripod into use for a much heavier camcorder. You're better served with a tripod that is more rugged than needed for the task, rather than having a lightweight unit tip over and damage your gear.

Stabilization Equipment

Some prosumer and higher-end camcorders include built-in optical stabilization circuitry, which automatically compensates for harsh or irregular camera movements to provide more fluid, more consistent movement. A number of third-part stabilization systems, such as those available from Glidecam (*www.glidecam.com*), increase the smoothness of camera movement further. Glidecam produces models that offer stabilization when a camcorder is being handheld, as well as small cranes, such as the Camcrane 200, that allow the camera to be smoothly moved through a wide range of motion. These types of systems range in price anywhere from $350 to over $19,000. Renting a system of this type can be a good way to determine whether it is something worth purchasing to improve the quality and consistency of your video work.

Other Gear

The range of camera mounts and related equipment can be very extensive, from mounts and fixtures that attach a camera to the bumper of a moving vehicle to robot-controlled arms that can execute smooth maneuvers under remote control from a sound-proofed control room (as discussed in Chapter 10). If you have very specific needs for gear that you may only need for two or three scenes, or your production is attempting to push the envelope and produce startling special effects, consult an experienced equipment rental service. Because you will typically be renting equipment that may costs hundreds or thousands of dollars, you will need to either demonstrate some form of insurance, provide a credit card with sufficient balance to cover possible loss, or put a sum of cash on deposit in an amount equal to the equipment cost.

As part of the pre-production planning, talk to local facilities and determine what their policies and rental costs are for the equipment you need. If you schedule use well ahead of time, you may be able to get a discounted rate by reserving the equipment for a particular period when demand is low. If budget constraints will be an ongoing concern during production, reserving and renting equipment during slow periods may be able to save you substantial amounts of money.

If you need very elaborate equipment to accomplish a particular scene, such as a large crane and cherry picker to accommodate a camera and camera person, schedule use of this equipment well in advance of production. Specialized equipment of this type may be difficult to find in less metropolitan areas and this type of gear may be reserved for weeks or months in advance.

Shooting Video Tests

Long before the full cast and extended crew show up on location for an actual shoot, you should perform some video tests at each site to be included in the video shoot. This approach, of course, won't work if you're videotaping a live play and you don't have access to the stage lighting equipment to simulate the setting that will exist on the day of the performance. But, in general, it helps to have as much familiarity with the locations involved in a shoot as you can. Consider ambient noise, activity in surrounding areas, lighting changes during different parts of the day, and other potential disturbances that may affect the production (such as clouds of mosquitoes that seek out cast and crew, or harsh afternoon sunlight that bakes everyone in a closed courtyard). Bring along those members of your production crew who may be able to help you evaluate the setting from an audio and video perspective. Try adjusting your white balance settings with the camcorder to see if you can achieve the lighting effect you desire. In general, you should try to spend at least a couple of days performing test shoots at a selected location so that you and the crew are thoroughly familiar with the conditions under which you will be working.

Another benefit of this approach is that you can perform specific tests using the equipment that you plan to use for the actual video shoot. This gives you an opportunity to assure the a particular camera or a particular microphone will work as expected during live conditions. Will a standard DV camera suit your purposes, or should you opt for a high-definition model? Will the new brand of microphones you are using with a new windscreen adequately perform in gusty, late-afternoon conditions? Will your hand-held camera movements work well in a brightly lit setting with lots of contrasts in the colors and light intensity? Will you be able to set up the equipment where you would like it to be or will the job require a different mounting apparatus for the camera or a different vantage point for the camera person? As much as you think you understand the setting and the changing conditions, performing some video tests may reveal problems that you hadn't considered and save you time and money when you get down to production with cast and crew.

Handling the Lighting

No single element will bear as heavily on the quality of your video as the lighting. Too many less experienced videographers and producers rely on the capacity of the camcorder itself for adapting to a variety of lighting conditions. While this can be very effective for documentary work and fast-moving sports events or concerts, if you're doing a feature film or a documentary with interviews or a travel piece, you can significantly increase the visual impact of your production through control of the lighting.

Even in situations where there is abundant natural light, you may want to employ umbrella reflectors to direct available light to the faces of your subjects in the video, whether actors or interview subjects. Applying professional lighting techniques to a digital video project is a sure path to achieving the maximum quality that can be achieved with the medium. This extra measure of quality makes a substantial difference when you begin encoding the video in preparation for mastering a DVD. No amount of post-production can compensate for failures to get reasonably correct exposures during videotaping.

Using Three-Point Lighting

Three-point lighting is a technique refined through professional studio applications, by which a subject is illuminated from three different directions. Done properly, this eliminates unflattering shadows and provides clear definition of the subject. The primary light is considered the key light and it generally is used at full intensity to provide the majority of light cast on the subject. A fill light from a roughly 90-degree angle to the key light softens the shadows and provides additional definition. A back light behind the subject is used to provide a consistent backdrop for the image, as shown in Figure 4 - 1.

Figure 4 - 1 **Three-point lighting**

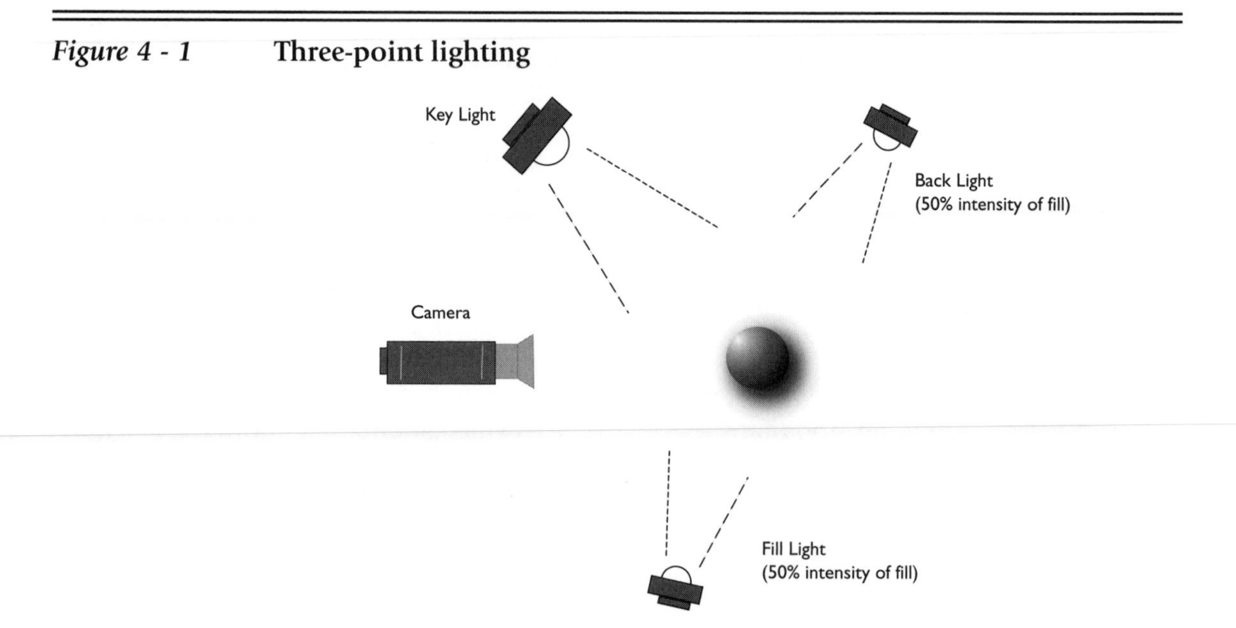

The single camera-mounted lights that are sometimes provided as an accessory for camcorders rarely work well in professional applications, since light directed squarely into the subject's face is not generally flattering.

If shooting in an outdoor location, try to schedule shoots for mid-morning or mid-afternoon to avoid the harsh shadows that can result when the sun is directly overhead. Even when using three-point lighting, overhead sunlight can interfere with your lighting and add harsh edges to the images. In comparison, shooting on cloudy days can often improve image quality by delivering more diffused light with softer shadows. When shooting in sunlight, you can sometimes use a reflector or white posterboard to illuminate the subject and to serve as the key light.

Color Temperature

Natural and man-made sources of light exhibit different color temperatures. Color temperatures vary according to what part of the spectrum is emphasized by the lighting source. We typically use the terms cool and warm to differentiate between the characteristics of light sources. Warm light sources (such as scenes shot in natural sunlight in early morning or late afternoon, or scenes lit by incandescent bulbs) favor orange and red tones and tend towards more saturated colors. Cool light sources include fluorescent lighting where the green and blue portions of the spectrum are favored and overall color saturation is less. A number of techniques exist to compensate for the color temperature to achieve a specific effect on a recorded videotape. Filters used over the camera lens, alternate light sources, white balance adjustments, and other techniques can turn warm light cool, cool light warm, or achieve a neutral balance. Many prosumer and professional camcorders feature a built-in neutral density filter than can be applied when the overall lighting is too bright to achieve the correct exposure even with the camera iris stopped down to a pinpoint.

Paying attention to the color temperature that prevails in a setting and making any necessary adjustments when videotaping gives you more precise control over the colors, as well as the mood, that prevail in a videotaped scene. Consistency is important so that you don't have too competing color temperatures within an image. For example, if a window is being used as the key light in a scene and the shadows are being filled with incandescent light from the other side of the room, the subject may appear pale and washed out on the window side of the scene and excessively flushed on the opposite side. Sticking with a single color temperature throughout a scene, usually by controlling the light sources, generally ensures the best resulting images.

Adjusting the White Balance

Most camcorders include some type of white balance adjustment. On less expensive cameras, this adjustment is automatic, either performed when the camera is initially powered on or by pressing a button.

Other higher end camcorders offer more flexibility when dealing with the white balance, which can be exploited to achieve a particular type of lighting within a scene. For example, the standard means of setting the white balance is to place a white card in front of the camcorder while activating the balance adjustment. If you substitute a tinted card, the entire spectrum of color as interpreted by the camera is shifted. This effect can be used to your advantage in certain kinds of scenes, perhaps to simulate a gloomy, twilight setting or to gain a pale, bleached look akin to pending snow blindness on a snow-covered mountain.

Onsite Considerations

As your site selections for a video shoot are made and your crew assembled, certain practical considerations should not be overlooked. Obtaining liability insurance to protect yourself, cast, and crew against unforeseen events may keep you out bankruptcy court. In most locales you will also need some time of permit to bring your crew in to shoot. Having heavy lighting equipment, electrical cables, and people blocking sidewalks and streets can create hazards; you need legal permission in most cases to perform these activities.

Getting Permits

Whether you plan to shoot your epic in a public park in the metropolis of New York or a lightly populated area in downtown Santa Fe, before you start actual work consult whatever city and state agencies are necessary to authorize your activities. Depending on the locale, you may need to pay a fee (usually a modest one in less-populated locations) and present proof of liability insurance. Research the necessary permit requirements well ahead of schedule to make sure your plans don't interfere with other plans in the locality (such as a street fair taking place the same day as your proposed shoot).

Permits can grant you access to areas where moviemaking opportunities abound: public parks, train and subway stations, shopping malls, outdoor stadiums. They sometimes include amenities, such as private parking areas and the ability to temporarily cordon off a section of street while the shoot is taking place. You should have at least one person on your staff whose job it is to ensure protection of the public (both those walking and those driving) from your moviemaking operations. In some cases, you may want to enlist the services of the local police to ensure a higher degree of safety for anyone venturing near the site. Such an arrangement usually requires paying the typical hourly rate for police services.

If your shooting plans change day by day, you should be cautious when reserving locations, particularly if expensive fees are associated with a permit. Many states have agencies set up to help moviemakers make connections with city authorities to obtain permits and select the best locations for shooting. For example, the Vermont Film Commission works with national and international filmmakers to help them choose locations, find equipment and specialists for their projects, and circumvent any obstacles that might discourage filmmakers from working in the area. Clearly, moviemaking crews can bring a good deal of revenues into a community, as well as providing some visual public relations in many cases. If you can align yourself with those agencies that want to ensure that your project is as successful as possible, you may gain some contacts and benefits that can greatly improve your experience.

Playing It Safe with Insurance

The most rigorous planning cannot eliminate the potential for accidents and the nature of moviemaking can sometimes invite accidents to happen. The work environment typically contains high-voltage electrical equipment, large numbers of people moving gear in every direction, locations where the public or traffic may get in the way, and so on. A single incident might be sufficient to shut down your entire project, so the safety net offered by insurance is not something to take lightly. Insurance may also be required to obtain permits in certain locales—check with the respective agencies to see what their requirements are and to ensure you have adequate coverage.

Among the forms of insurance that should be considered for any moviemaking operation are general liability insurance, worker's compensation insurance, and equipment coverage policies to protect you against the loss or damage of expensive gear. Insurance coverage differs widely from region to region. Some policies are designed to cover only a particular shoot. Others offer coverage on an annual basis. Even insurance companies that are not necessarily set up to service moviemaking production companies can sometimes provide general liability coverage for your business. A camera and equipment rental service in your area may be able to offer a referral for an insurance company specializing in your needs. Shop around on the Internet to try to get some idea of the rates associated with the coverage you want. A liability policy that provides reasonable coverage for a small to medium-sized crew may carry an annual premium in the range of $600 to $1200. If there is even a single event where the insurance can help you avoid a much larger loss, this level of coverage can be a bargain.

Monitoring and Capturing Sound

Sound is often a neglected commodity in audio-visual productions, whether you're talking about documentaries, film, or video work. This is unfortunate because in general most audiences are much more forgiving about lapses of quality in images onscreen than they are about audio quality.

Many sound engineers and production sound mixers agree on one central point: if you want the best possible sound in a production, capture it well in the field. As with video, there is much you can do in the post-production phase to filter, sweeten, massage, and smooth out the audio. You can even put the actors back into the studio and have them lip sync the dialogue to get it exactly right (a process known as Automatic Dialogue

Replacement—ADR). However, for efficiency, cost savings, and, ultimately, more pleasing results, you should do whatever is needed to record the highest quality audio possible while in the field.

Unless you happen to have expertise in this area, a talented production sound mixer can make a world of difference. Typically, this person bears the responsibility for selecting microphones to be used, determining the optimal microphone placement, directing the boom operator when appropriate, and ensuring the proper levels and mix while shooting.

Recording the Audio

Independents working with digital video camcorders often vacillate over whether the audio should be captured on camera or with an external recording device, such as a digital audio tape (DAT) recorder. While a DAT recorder can be an excellent tool for picking up ambient sound, sound effects for a production, narration or dialogue that needs to be repeated, and even live music, there is a clear benefit to recording the audio directly into the camcorder. Most DV camcorders can record stereo audio tracks using 16-bit samples and 48KHz sampling rates, but the audio circuitry available on many DV camcorders can be lesser quality than a professional quality DAT recorder. On the other hand, if you record the soundtrack separately to DAT, you lose synchronization with the video and must resync during post-production. This can seem like a minor matter given the flexibility of non-linear editing systems, but it can be a time consuming and grueling process.

The other frequently made mistake is to rely entirely on the built-in microphone included with the camera. Audio levels are either preset before a scene or captured with Automatic Gain Control (AGC) active. AGC can be handy for doing documentaries where sound levels change frequently, or for other applications, but this feature lacks the precision of doing live sound mixing to best capture the audio on location. AGC tends to bump up the volume for quieter passages, whereas in the scene you may want a section of relative silence without the rustling of foliage, the chirring of crickets, or the rushing of the wind.

For many productions, the best solution may be to use external microphones run into a sound mixer, which is monitored and adjusted by a skilled professional as the scene is shoot. The audio can then be routed from the mixer to the camera, providing the necessary synchronization to dramatically cut down on the post-production work. You may still want to sweeten the sound quality and add effects during post production, but, with some care, you eliminate the need for the actors to have to redo any of their dialogue weeks (or months) later, a tedious and expensive process. This alone is one very strong motivation for getting the sound well-recorded in the field.

Guidelines for Creating Exceptional Audio

These tips and guidelines can help you get the best results when recording audio in the field:

- **Think through the audio before the shoot begins**: Ambient sound can change dramatically during the course of a day, particularly in city locations. When doing video tests or scouting locations, pay attention to the sound atmosphere. Will the late-afternoon commute traffic ruin the audio for your establishing shot of a historic building? Is a construction crew beginning work one street over? Would it be better to record during a different time of day? Let your sound mixer analyze and evaluate the locations to be aware of potential audio problems before they occur.
- **Consult with the post-production sound facility before shooting**: Having a strong understanding of the requirements of a post-production sound facility that you have selected can help eliminate potential problems from developing in the field. Talk over your project with the professionals and exchange ideas to be sure that both sides have a clear perspective on how the audio will be handled. Budgets can be busted by overlooking simple points that might easily be highlighted ahead of time by having this kind of discussion.
- **Don't skimp on the microphones**: Despite the myriad of technical advances in computer and DVD technology, seasoned veterans in film and television often rely on proven microphones, such as the Sennheiser 416 and some of the Neumanns and Schoeps models. A predictable unidirectional pattern is essential if you want to minimize the chances of picking up off-axis sounds. Some of the new digital wireless microphones are gaining in popularity and these are worth considering for specialized situations. The Lectrosonics UCR411 and UM400 have won respect from veteran sound mixers.
- **Don't employ different types of microphones during a single scene**: The audio produced by particular microphones tends to be quite distinct and if you switch from one type to another during a scene, the audio effect can be jarring. For example, going from a

boom-mounted unidirectional mic for part of the dialogue and then switching to a wireless lavaliere mic later in the scene causes the audio to sound sharply different.

- **Open up your options with multi-channel recording**: If you decide to record to a separate source from your DV camcorder or if you're shooting film, you might consider one of the capable multi-channel recorders available. The Fostex PD-6 DVD recorder provides six-channel field recording, allowing six different mikes to be positioned about the scene. For projects where the ultimate audio track is going to be mixed down to 5.1 Dolby Digital format, these six channels provide a broad spectrum of sound and dimension for the sound mixer to use. The complexities of performing multi-track audio recording in the field have discouraged this type of approach in the past, but new tools are providing new opportunities.

- **Record two or three minutes of ambient sound at each location**: During post-production, it can be helpful to have a sequence of recorded "silence" to insert into the video where other audio is cut. There is a characteristic background sound even in moments of silence in a scene and this can be difficult to recreate (and it can sound very odd when it is not present). Having a segment of ambient audio to splice in where needed may smooth out the audio editing.

- **Record a "wild track" if the sound mixer detects problems during recording**: A wild track consists of audio (usually dialogue) that is recorded after the completion of a scene to correct a problem that the sound mixer detected during recording. Having this wild track available during post-production can often avoid the need to bring the actor back into the studio to re-record dialogue; think of it as an audio insurance policy.

- **Listen to sound being recorded**: One of the benefits of DV moviemaking is the immediacy of the process. Take advantage of the feedback that you can get in the field by stopping once or twice a day and reviewing the audio and video (assuming you're recording the audio directly to the camcorder). Usually the sound mixer will pick up problems as they happen, but sometimes having cast and crew evaluate a segment of video can reveal areas where improvements can be made. These kinds of mid-course corrections can often boost the overall results of a shoot.

5

Handling Assets: Managing Content for a DVD Project

The assets incorporated into a single DVD project can number in the hundreds, encompassing still photographs, MPEG-2 video segments, audio in a variety of formats, graphics for menus and backgrounds, and so on. Without some system for organizing and coordinating these assets, you could waste a considerable amount of project time just searching for items that you need. To deal with the massive volumes of files that go into any project, you may want to devise a system to track, view, and identify files. The system you design should be flexible enough to search for items by different methods of organization (such as dates or keywords), but simple enough so that you don't have to devote weeks to creating a database for asset management.

The prospect of creating elaborate naming systems, cataloging endless collections of sounds and video clips, sorting and filing enormous libraries of digital assets, doesn't much fit the image of a leading edge DVD producer leveraging space-age tools to dazzle a world-wide audience. This mundane work, however, puts you in a much better position to deliver top-notch creative and artistic content. If you don't know where to find that stunning audio clip of the South American tree frogs shrilling in the night, you can't embed it in your production. If the colorful, chaotic scenes shot during the last Mardi Gras have vanished from your reach, you might as well have never shot them.

This chapter explores some of the ways you can better organize and access content for a project.

Basic Premise: Keep It Digital

The range of assets for any possible DVD project can run the gamut from the most analog of analog sources (unretouched photographs from the Civil War era) to the most digital of digital sources (uncompressed digital video content in voluminous quantities). As a general rule while working on a DVD project, when you are archiving and creating asset libraries, try to keep all original assets in the digital realm in the least compressed format possible. Video in the DV format undergoes 5:1 compression when initially recorded to tape and then it is compressed again in MPEG-2 format for inclusion on a DVD. Having a version of this content in the least-compressed form possible gives you the opportunity to return to original shots whenever necessary so that you can reuse the content as required.

The Evils of Generational Loss

If you're working with old black and white film footage from the 30's, once it undergoes the telecine process and you have it in digital form, store and archive it in digital form. Large volumes of content of this sort may require high-volume solutions, such as digital linear tape with capacities up to 120GB, for reasonable storage. Smaller digital file sizes can be more conveniently archived on recordable DVD, DVD-RAM, or even CD-ROM. The basic idea is that when working in the digital world you want to avoid the generational losses of going from analog-to-digital or from digital-to-analog. If your input sources or output devices are, of necessity, analog in nature, try to limit the conversion to the single step of inputting or outputting, as necessary. Digital assets can be copied, manipulated, edited, transferred, or otherwise handled without incurring any type of degradation in the original source material. In comparison, the analog world offers a multitude of progressive content degradations each time you make a copy of a tape or each time you convert from one format to another. If you can stay in the digital realm for the majority of your production process, the overall quality of your production will be significantly cleaner.

DVD is, by nature, a vehicle for delivering digital audio, video, and stills. Think in digital terms. Stay in the digital realm whenever possible.

The Evils of Compression

One of the downsides of the digital realm is compression. To adequately store and transfer the large volumes of image content involved in digital motion pictures, some form of compression is necessary. The basic digital video images stored in DV format are captured to videotape using a 5:1

compression ratio (using Discrete Cosine Transform—DCT—algorithms, which are closely related to MPEG-2 compression techniques). Some of the more professional digital video formats use lesser levels of compression, but most formats require that the bit patterns get transferred and stored using some form of compression.

The DVD, of course, carries the compression one step further, using MPEG-2 storage techniques to produce video that is sufficiently compact to be delivered to players at a data rate that won't overwhelm current state-of-the-art electronics.

Unless you are dealing with a lossless compression format, you are always dealing with an image that is somewhat reduced from its original quality, regardless of whether that image was captured with a pristine Leica lens and digitized using three-quarter-inch CCDs. Once digital information is lost, it's gone for all time. For that reason, when storing and archiving content for projects, use the least compressed storage format available.

For standard digital video content, the ideal storage format might be the originally imported files in QuickTime or AVI format. For an imported still photograph, that might be the original scanned version captured at a high resolution (before being scaled for inclusion in a DVD project). This principle results in larger files for storage, but also ensures that you aren't storing any content at less than optimal quality. DVD presents video that is significantly better than any previous home-oriented format, but that video is streamlined through the compression process and reduced in size as much as possible to simply fit on a disc. An archive should contain the highest-quality version of the source content in digital form. Follow this principle rigorously and you'll never have to compromise your production by working with a lower quality source file.

Storage Strategies

As an overall strategy, organize the content that you need according to how frequently you need to access it. Typically, content can be organized into three basic categories, ranging from easy to access to more difficult to access:

Online: Assets that can be accessed either from local storage repositories or available network resources are the easiest to work into a project. Depending on the size and complexity of your local network (or individual computer linked to the Internet), these kinds of assets might include stock video stored on a RAID system, a library of photographs on a network disk drive, or a collection of audio stingers organized on your personal computer. Obviously, online assets are the most accessible, and, if you know what you're looking for, the easiest to find.

Nearline: Nearline storage includes those assets that you have occasional reason to access. These can be stored on a number of different types of media, such as CD-ROMs, DVD-ROMs, DVD-RAMs, Zip disks of various capacities, portable hard disk drives, optical disc juke boxes, and so on. Nearline storage, as opposed to offline storage, should provide a rapid, non-linear means of obtaining required files (which rules out backup tape drives and videotape cassettes, where all data is stored in a linear sequence).

Offline: The term offline storage generally applies to media that needs to be accessed through a linear method, such as getting files off a backup tape drive, or converted in some way from original source materials, such as using a digital capture board to re-import analog video content. Offline storage methods require the most time and effort to gain access to files. These types of storage devices, however, can be useful by providing an inexpensive mechanism for archiving extremely large volumes of data.

Using Project Storage Effectively

Non-linear editors often use the metaphor of the *bin*, much as the computer *desktop* have become the prevailing metaphor for an individual's workspace. You can assemble the assets required to build a video sequence or presentation by identifying the contents in a series of bins. Generally, bins contain pointers to the original source material, rather than actually moving files into new storage locations. This technique preserves disk space while providing the illusion that you are filling individual bins with the files you need to create a project.

As a complement to this approach, many non-linear editors and audio editing tools apply a technique known as non-destructive editing. This technique maintains the assets for a project in their original state, but stores the changes that you specify to them in a project file. All the video cuts and added transitions and audio splices are handled through a set of instructions that can be applied to the original assets. While you can typically preview the results of your editing, the source content remains unchanged and undamaged. If you change your mind about an edit, it is a simple process to go back to the original file, which resides in its storage location untouched.

Some applications, such as Vegas Video 4, can automatically add any captured media, such as video clips accessed from a camcorder, to a Media Pool area. If the original file cannot be located later, Vegas Video gives it a Media Offline label and gives you the option to recapture it from the camcorder or video deck. Other applications have similar capabilities.

The method of organization that you decide you use depends entirely on your personal preferences. Many NLEs, such as Final Cut Pro and Vegas Video, give you the option to nest bins inside other bins, creating a hierarchical structure. You can establish a flat structure, with lots of bins at the top level, but little nesting underneath (which avoids having to dig through the hierarchy to find things). Or, you can create a deep structure, which may only have a few bins at the top, but a series of progressively deeper levels that contain bins organized in a specific way. For example, you might have a top-level bin labeled Background Audio which contain several bins with categories such as Country, New Age, Jazz, Salsa, Reggae, and so on. Inside each of these bins might be another series of bins with the length of the audio clips: 30 Seconds, 60 seconds, 90 seconds, and so on. A deep hierarchy lets you provide a more elaborate organization, but unless someone understands how the bins are organized, it can be more difficult to find things. In comparison, a flat structure gives you an immediate broad overview of the full range of content, but the top level can get crowded and confusing if too many bins are added. The structure that you use depends largely on how you personally organize and think about the content. If you work with a project team, make sure that everyone understands the chosen structure.

As a general rule, move the assets that you intend to use in a project to an online location, preferably to a fast hard disk drive that is dedicated to your project. If assets reside on lower-speed media, nearline storage resources that require excessive access time, or offline storage, when it comes time to render the audio and video content, the process may be slowed or interrupted by the slow access speeds or temporarily inaccessible storage devices. A large-capacity SCSI or IEEE-1394 drive can generally provide the best results and performance for asset storage and access.

Consider these guidelines for organizing project assets:

- Many of the NLE applications include features that simplify asset management and organization. For example, Adobe Premiere lets you add descriptive comments and keywords for each element stored in a bin. Learn to use the available application tools for fully and accurately describing your assets (and also accelerating searches); when your project expands to hundreds of assets, you'll be very glad to have these descriptions to clarify the details.

- The bin concept of storing assets consists of pointers, which lets you store the assets anywhere, but for simplified management and location of key content, you might want to assign a dedicated folder and series of sub-folders on a high-performance hard disk drive where you routinely capture incoming video, copy assets produced in other applications (such as MIDI files or Adobe Photoshop graphics), and render interim versions of content. Having this type of structure makes it easier to manipulate and view content when you are working outside of an editing application, and also makes it much more convenient if you want to perform large-scale backups of assets. Bins and pointers are helpful, but you can complement this approach by consolidating assets in a single area.
- If you often have a number of different projects going on at the same time, you might want to consider purchasing an application that helps sort and preview assets. A number of different applications exist for managing assets and resources, as described in the next section. These can make a big difference in selecting content to use within a project, as well as bringing some sanity to organizing the assets in large-scale productions.

Asset Management Applications

While video editing applications often include internal methods for storing and accessing assets, you can gain a more global perspective on your content and asset libraries through dedicated programs that categorize, sort, preview, and streamline access to assets.

Extensis Portfolio

Extensis Portfolio 6 represents one of the more intuitive applications for handling media content, with versions available for both Macintosh and Windows systems. Portfolio automates many of the common asset management operations, letting you create folders and subfolders that are automatically monitored for new files. As files are dumped into folders from other applications (such as when you create graphics in Adobe Illustrator or Photoshop), Portfolio can add them to your library where they will be displayed the next time you open Portfolio.

A system of floating palettes gives developers a quick technique for pulling cataloged files out of libraries and dropping them into an application—such as when a developer is selecting assets to use in an NLE application. These palettes, available in Portfolio Express, float beside the media bins in the application in which you are working and enable you to simply drag and drop content where you want it.

For project teams that may be dispersed over a wide geographic area, Portfolio makes it easy to share catalogs of assets. A catalog can be created dynamically, moving all the collected files to a new location, such as a recordable CD or DVD-R disc. Thumbnails and compressed previews of asset files can also be displayed through posting to the Web, with HTML pages being created based on a number of pre-set templates. If DVD project teams are working on a variety of different assets from different locations, these Web postings can help the collaboration effort and keep everyone on task.

See *Audio Work in Vancouver* on page 209 for an example of how Internet collaboration can improve the quality of a production.

Portfolio can handle the raw video content composing a DVD project in either AVI or QuickTime format, but version 6 does not have a provision for storing and accessing MPEG-2 video files. An extensive variety of other files useful in DVD work can be cataloged and accessed, including:

- Adobe Photoshop v4.0 - 7.0
- AIFF sound files
- Kodak PhotoCD
- Common graphics formats such as GIF, JPEG, PNG, TIFF, BMP and PICT
- MIDI sound files
- PostScript
- Text
- WAV sound files
- Macromedia Flash and Shockwave
- Raw Digital Camera format

The support for Pro Photo Raw image files provides direct access to the content from high-end digital cameras available from Fuji, Canon, Olympus, and Nikon. These uncompressed images can provide exceptional source content for DVD projects when rendered to appropriate resolutions and added to sequences.

More details on Extensis Portfolio can be found at: *www.extensis.com*.

Capitalizing on Asset Management

Developing a workable asset management system for your DVD projects delivers benefits that far outweigh the time and effort required to set up the system in the first place. With each DVD project that you complete, you will save time and be able to make more effective editing decisions by having a wider view of available assets. Just the time and frustration saved in hunting for lost files and vaguely remembered sequences will probably make the setup process worthwhile. You don't have to develop a form of asset management to work on DVD projects, but if you do, it's unlikely that you'll ever want to work with a chaotic, unorganized library of assets again.

6

Creating a Story: Using Non-Linear Editing Tools

Among its many capabilities, DVD provides an effective medium for telling stories—digital video has become a driving force in creating content for distribution on disc. Whether your story is a fictional tale set in a fully-realized world in the 23rd century or a documentary on the impact of blues music on American culture, digital video offers a number of advantages over other techniques for assembling images and sound into a presentation. The ease at which multiple ideas can be assembled, examined, layered, and re-assembled easily surpasses any previous form of editing. The fluid nature of this process using non-linear editors offers unprecedented freedom and the ability to experiment easily and inexpensively, a crucial part of any genuinely creative process.

Given the nature of current applications, many of the tools and processes have been designed to be interconnected. On the front end, a generation of prosumer DV camcorders provide a direct means of capturing recorded DV content to the computer through FireWire/IEEE-1394 connections. Note that many professional DV cameras still don't provide this direct link and require that DV be transferred from tape to computer in a much more circuitous route. On the back end, many NLE applications provide considerable flexibility in outputting digital video content to a variety of compressed and uncompressed formats. Some tools, such as Apple Final Cut Pro and DVD Studio Pro, include features to simplify chapter creation for DVD. Chapter markers specified in Final Cut Pro can be read and then incorporated in the disc image created by DVD Studio Pro, adding much efficiency to the process of making DVDs. This chapter explores the features and capabilities of non-linear editing tools and provides tips and guidelines for using these tools in your projects.

The Desktop Production Studio

The functionality within a modern non-linear editing application spans a variety of different specialties. In comparison, analog video or film editing suites often require additional (and expensive) components to achieve the same functionality. For example, the capabilities of Adobe Premiere Pro, the current generation descendent of the most successful of the original computer-based editors, include video editor, special effects generator, titler, and compositor. Programs such as Adobe After Effects open up the full realm of digital image manipulation, creating effects, transitions, animation, and video processing techniques that are limited only by the imagination of the content developer. With the booming popularity of digital video tools, the feature sets of even modest applications are surprisingly robust. For example, MainConcept, the company that licenses the MPEG-2 encoder used in many popular NLE applications, markets their own video editing application, MainActor, with a feature set comparable to products that cost hundreds of dollars more.

The basic organization of most NLE applications has evolved to a common model that evokes earlier generations of production tools. The features common to most NLEs include:

- **The Bin**: Serves as the storage receptacle for video content and other elements to be used in a production, such as music clips, stills, and so on.
- **The Monitor**: Provides a viewing window for footage, which can be used to mark points in a video sequence.
- **The Timeline**: Offers a quick and convenient means of ordering and organizing the content of a production that can be collapsed and expanded as needed.

These elements can be traced to the origins of film editing, where an editor arranged sequences of film from strips suspended within arm's reach. Physically splicing together the sequence of clips, the editor could then view the trimmed and assembled results on a viewer centrally positioned on the editing workbench. As shown in Figure 6 - 1, in a typical NLE application, a large amount of information can be presented in a manner that makes it very easy for an editor to work.

Figure 6 - 1 **Editing Interface to Sony Vegas 4.0**

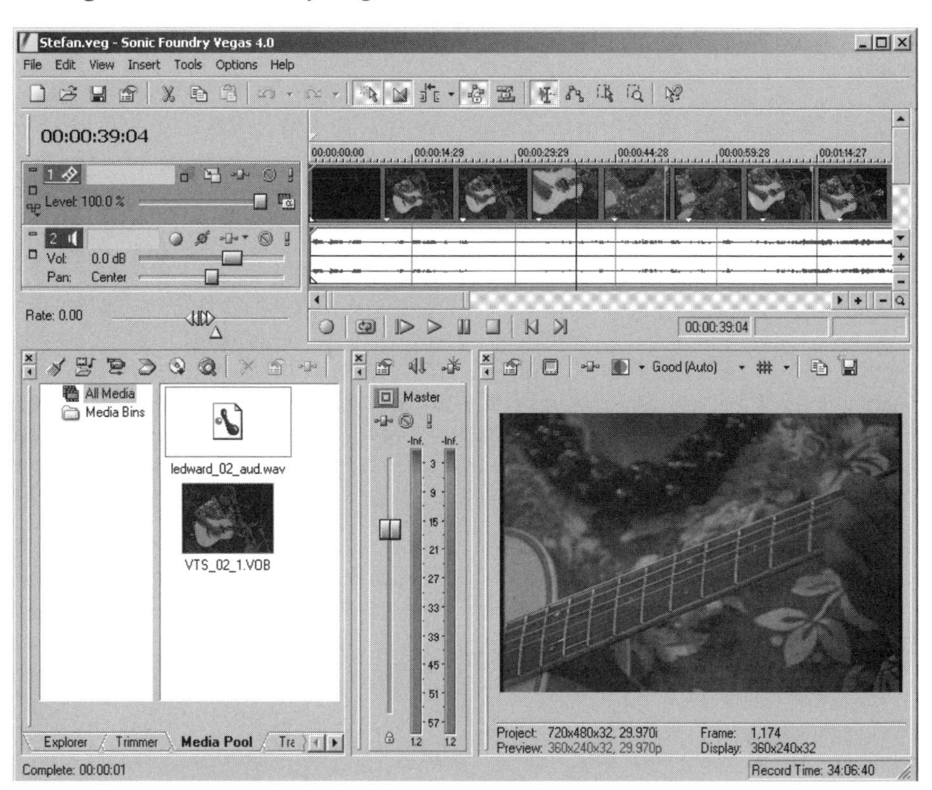

For projects where the end goal is creating a DVD-Video disc, the editor works within the NLE to create one or more video sequences. The resulting video can often be directly encoded to MPEG-2 format for use with a DVD authoring package. This process also multiplexes the audio and video streams into a format that is appropriate for DVD presentation.

Managing the Capture Process

Both analog and digital video content can be captured to an NLE application, a process sometimes called logging. From an excess of raw video footage, the goal is to pare the source material down to a manageable level and also to devise a method of labeling and identifying the material so that it can be handled more effectively during editing. The logging and capture process can seem tedious and pedestrian, but if done effectively, it becomes much easier to work with the available content and produce worthwhile end results.

Regardless of your type of project—documentary, feature film, or marketing presentation—you will invariably have shot more content than you need. You can simplify and streamline editing work by judicious trimming of unnecessary content during capture. For every one minute of completed video, you may have shot ten minutes of actual footage (or more). If you can pare this down to a 5:1 or 3:1 ratio for editing within the NLE, you will save on storage space and streamline the editing.

Digital video captures are fairly straightforward. Using a batch capture technique, you can often use your NLE application to control the video camcorder or deck attached to your computer system. Then, you can progress through the video content, marking the in and out points of the segments that you want to transfer from tape to hard disk. Once you've made these decisions, you can initiate the batch capture and the software transfers the specified content to the indicated disk drive. You can often set up naming conventions so that individual segments will be tagged with identifiers as they are automatically captured.

Figure 6 - 2 shows the capture window as it appears in Apple Final Cut Express. Note that the left side of the window provides a view of the content on the videotape, which can be scanned and examined using controls similar to a conventional video cassette recorder. A shuttle control lets you precisely advance and rewind the tape to select in and out points.

The right side of the window gives you the opportunity to label the captured video using a number of fields. The more care you take during this part of the process, the easier it will be to initiate searches when your project becomes more complex. You can employ three different capture methods—Clip, Now, or Project—to import data into your project.

Figure 6 - 2 **Capture window in Final Cut Express**

Alternatively, you can grab individual segments from a videotape as you need them. Most applications support a function where you can click a capture or record button while viewing a videotape to begin transferring the content to disk. When you have as much of the segment as you need, you can simply select stop.

The process of culling down the content to meet editing requirements can shape the structure of the entire work. You can always go back and get more content from videotape as needed, but you may be better served by marking your in and out points for capture carefully on the first past. As you work with material during editing, you don't want useful or interesting segments forgotten on your source tape.

Given the flexibility of digital video editors, some content developers prefer to input the whole range of material with which they will be working, without attempting to discriminate between useful footage and useless material. This technique can be used if you have an abundance of hard disk storage space (preferably more than 200GB) and prefer to view and cull material on the computer, rather than on the camcorder or deck. If you are working exclusively from a digital video camcorder (rather than a video deck), you may want to capture entire videotapes at once just to save the wear and tear on your camcorder tape drive mechanism. Starting and stopping and rewinding can be intensive when you're trying to mark in and out points and this activity contributes to the wear on your camcorder.

Analog video capture methods usually offer more flexibility in terms of specifying the data rates (and thereby file sizes) when capturing. The degree of control over the camcorder or deck may be less than what you have available for DV camcorders. You may need to manually start and stop the videotape as you use the capture or record function in the NLE.

Timecodes

Working with video editing tools, some means of clearly and precisely identifying video segments is necessary, from videotape input to final production output. The standard, embraced throughout the television broadcast industry and beyond, is SMPTE, a standard defined by the Society of Motion Picture and Television Engineers. SMPTE encodes a timing signal on the address track of a videotape which can then be imported and used for reference when editing.

Two variations of the timecode presented by SMPTE exist for NTSC content, which is often a source of confusion, since many NLE applications accept and work with both formats. A single time code works for PAL content.

- **NTSC SMPTE Non-Drop Frame**: NTSC video runs at 29.97 frames per second with each frame composed of two interlaced fields clocked at 60Hz.
- **NTSC SMPTE Drop Frame**: For broadcast purposes, the NTSC signal is actually 59.94Hz, which corresponds with a frame rate of 29.97 frames per second. To compensate for this small adjustment, a frame is dropped roughly every 33 seconds. By this technique, an hour long broadcast fills exactly 60 minutes, rather than coming up a few seconds short as it would using SMPTE non-drop frame format.
- **PAL SMPTE**: The PAL format, as used by the European Broadcasting Union (EBU), is much simpler since the values are more evenly divisible. A 50Hz clock regulates frame display at 25 frames per second. Thus, no drop codes are necessary to adjust for mathematical variations.

Mixing non-drop frame and drop frame source and output formats can cause a variety of problems. For example, synchronization between audio and video tracks can be lost so that before long the dialogue doesn't sync with the onscreen character's lips. Stick with one format or the other throughout your video editing process. If you want to work with PAL content using an NTSC timecode, you need to rerender the video—simply selecting a new timecode for your video project won't do the trick.

Reels and Timecodes

A common technique for handling project content is to set the starting timecode on the camcorder to correspond with the reel (videotape) in use. For example, it you're beginning your third one-hour videotape cassette, use a starting time code of 03:00:00:00. A sequence of project timecodes such as this can be very helpful when importing the video into your NLE application. For convenience, you might create an individual bin for each reel, making it easier to sort and organize captured content.

Consumer and prosumer camcorders don't typically provide a method for setting a starting timecode value. For example, the Canon XL1 resets to 00:00:00:00 each time you insert a new videotape. You can, however, still direct the captured content into a bin number that corresponds with the hour of the segment from the tape.

If you are capturing the contents of an entire videotape, rather than individual marked clips, make sure that the timecode as recorded on the videotape is contiguous throughout the length of the recorded material. Breaks and gaps, or restarts of the timecode, can cause havoc when you begin editing digital video content, since the timecode is intended to synchronize and coordinate the project contents.

Offline Editing Based on Timecodes

Many of the major NLE packages offer the ability to generate an edit decision list (EDL), which is essentially a collection of all the editing cuts and transitions made throughout a project. The usual strategy is to work from a low-resolution video with a contiguous timecode and use the timecode to specify the starting and ending points for each cut. This type of editing activity is typically used when the source material will be edited on a non-digital system, such as when a 35mm film is physically cut and spliced per the editor's requirements or when an analog video editing system will be producing the final content. The contents of the EDL can be submitted to the film editor or video editor to guide the production work.

Composing the Story

By the time we have reached adulthood, we have all absorbed thousands of hours of visual content, spanning television, cinema, billboards, magazine graphics, creative photography, and so on. To a large degree, our culture has internalized many elements of this visual language and our viewing of visual content has become something of a meta-language. The moment you start creating sequences in a video editing application, this language comes into play and the success of your story depends on your using visual elements effectively in combination with the audio content of your production. The storyteller working in the digital video editing realm has an important advantage over analog video editors and film editors: the ability to experiment freely to achieve the exact effect that you want.

The immediacy of being able to preview your editing decisions makes it possible to change the tone and timbre of your story on the fly. If the pace is too languid, you can shorten clips, interpose more angles, substitute a more energetic audio track, overlay segments to create motion effects, and view the results either immediately or after a short rendering delay. Faster processors have brought us to the point where many complex video editing effects and transitions can be accomplished in real-time, bypassing the need for a lengthy rendering session to generate a preview.

As you become more engaged in developing video content, you may start analyzing television shows and films to see what techniques experienced directors and editors are using to achieve their goals. Casting such a critical eye on entertainment pieces may detract from the excitement or relaxation of the entertainment, but it can contribute significantly to your overall video editing education. Pay attention to those works that you particularly enjoy and try to understand what elements you find most appealing and consider how these effects were achieved. Entire books have been written about storytelling techniques in video and film, and this enormous topic transcends the space available in this book, but hands-on learning can be the most effective way to master any craft. Learn from the masters and then add your own personal style. Unlike the craft of filmmaking, digital video content can be created inexpensively and the resulting works distributed equally as inexpensively on DVD. This creative freedom in the visual storytelling medium is unprecedented and heralds a whole new era that may expand the boundaries of our communication.

Editing Example in Vegas 4.0

In the course of working on the case study about Edgewood Studios, I videotaped David Giancola giving a tour of the facility with the help of Edgewood staff members. As an example of how the digital video editing process works, this section covers the editing process and illustrates how a 45-minute segment of source material was edited down to a much shorter segment for inclusion on the DVD bundled with this book.

A Canon XL1 camcorder was used for the videotaping and the audio captured with a Samson AirLine UHF wireless microphone worn on the shirt of narrator David Giancola.

Capturing From Tape

Depending on your preferences, you can capture the entire contents of a videotape from deck or camcorder, or you can more selectively capture content by specifying mark in and mark out points while previewing the tape. This more selective process is effective only if your camcorder can be directly controlled by the video editing application, usually through a FireWire/IEEE-1394 link. If you have limited amounts of storage space available, you will probably want to be more selective about your capture technique.

As shown in Figure 6 - 3, the capture window in Vegas can be used to direct video files into the Media Pool. If you move or delete video segments that you need later, they can be recaptured from the original source videotape (be sure that you label the tape so that it corresponds with the identification that Vegas logs for it) using Recapture.

Because files included in a video project are referenced through pointers (rather than moving the physical file to a new location), you can organize the content for your project any way that is convenient. The only caveat is that files must be available when the project is rendered (if they are offline, you must perform a Recapture to continue).

Figure 6 - 3 **Video Capture Operations**

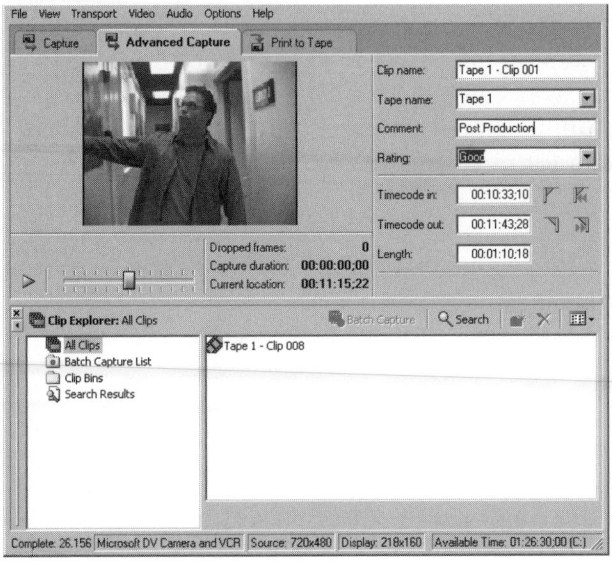

Media can also be added to the Media Pool from other sources, including audio, graphics animation files, still images, and so on. One advantage of Vegas is that it can seamlessly handle very different media types and sizes, and automatically convert them upon rendering to fit the project parameters. This saves time in scaling graphics, sampling audio, and so on, but care must be taken not to include media that will lose quality when converted. For example, the difference in pixel aspect ratios between digital video and conventional computer graphics requires consideration when creating images to bring into a production. To create full frame images corrected for the aspect ratio differences using an application such as Adobe Photoshop, use these dimensions:

- **NTSC DV**: Create still images at 655 x 480
- **PAL DV**: Create still images at 787 x 576

When placing audio content in the Media Pool, try to use sample rates equal to or higher than the desired output format. For example, DVD audio is by default two-channel, 48KHz, 16-bit samples. Vegas can upsample lower resolution audio, but the quality will not be as sharp as when a higher resolution sample is used initially.

Trimming a Clip

For every operation in Vegas, there are generally three or four different ways to perform it, ranging from keyboard shortcuts to drag-and-drop techniques. The contents of the Media Pool can be dragged onto the timeline, where they become *events*, distinct audio-visual elements, such as a video segment from a longer clip or an indicated audio clip. Non-destructive editing, being based on pointers, is extremely flexible in that you can perform many different operations and, if you're not satisfied with the results, you can employ multiple levels of undo to progressively back away from your changes. When you trim a video clip, whether it is in the Trimmer pane, or directly on the timeline, you are simply telling Vegas where to start and end that clip during rendering.

The following figure, Figure 6 - 4, shows the appearance of the Trimmer pane.

Figure 6 - 4 **Trimmer Pane**

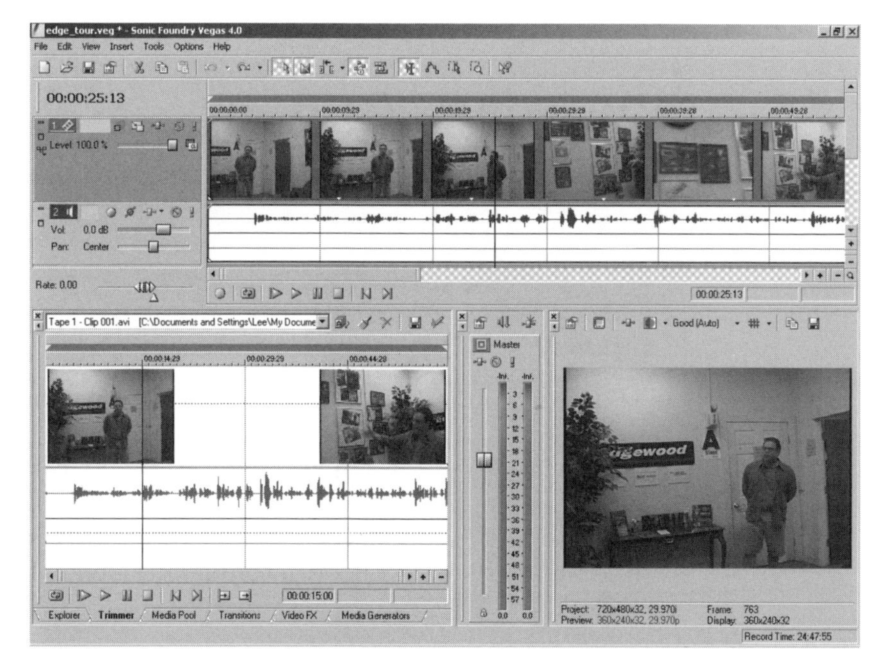

The Trimmer pane, in the bottom left quadrant of the screen, shows starting and ending frames, as currently selected. This trimmed sequence can be dragged to the timeline, or added through shortcut keys or insertion buttons. Trimming is only one way to modify the contents of a video or audio segment. Vegas supports a wide variety of cut and paste operations, including compositing of frame contents by overlaying segments on top of each other through cut and paste.

Creating a Title

Another of the advantages of the latest generation of digital video editing tools are the many built-in features that used to require separate applications. Titling is one area that has a wide variety of applications, ranging from creating basic title displays to designing rolling credits to generating closed captioning. Vegas includes robust titling options through the Text Media creation tool, as shown in Figure 6 - 5.

Figure 6 - 5 **Text Media Creation Tool**

The titling capabilities include full control over placement, font, size, alignment, and other properties, as well as the ability to add keyframes and animate the text. Smooth keyframe animation can be applied to effects such as text size, color, leading, tracking, and position. The completed element of text media can then be positioned at any point on the timeline and used as an overlay or standalone element.

Making Transitions

Transitions are abundant and versatile in Vegas. The temptation to experiment with this never-ending supply of built-in after effects can be considerable, but in actual practice, many video projects can be completed successfully using simple cuts and dissolves. Like font availability in word processing, video transitions work best when they are used with taste and discretion. If overused, they become tacky and inappropriate.

Figure 6 - 6 **Video Transition Options**

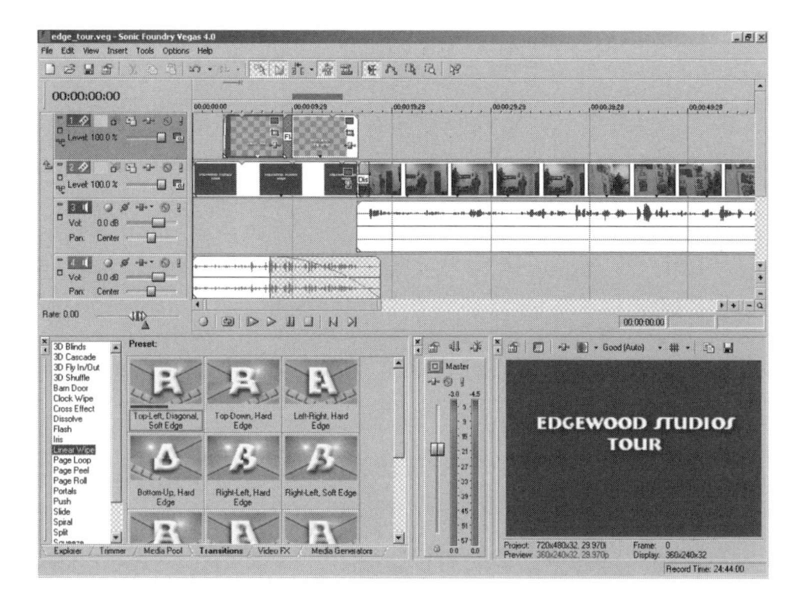

The Transitions pane, shown in the lower left portion of Figure 6 - 6, includes a small thumbnail that shows graphically the effect of the transition. Transitions can be applied simply by dragging them to the position on the timeline where the transition is to occur. You can infinitely tweak and adjust the individual settings for each transition to get precisely the effect that you want. On a single video track, the overlap between two adjacent video segments determines the period during which the transition takes place. You can easily shorten or extend the length of the transition by dragging one of the overlapped video clips forward or backward along the timeline.

Adding Markers

Markers indicate specific points along the Vegas timeline that can correspond with timing cues, navigation points, and, when exported to DVD Architect, chapter starting points. Vegas supports a number of different types of markers, including command markers that can be employed to created closed captioning for Windows Media Video files. For DVD creation, however, using markers to designate chapter breaks can save a bit of time when you go to DVD Architect to perform authoring. This operation can also be carried out from within DVD Architect, so completing the task while in Vegas is not essential, but, if nothing else, it can help organize your video presentation into precise units.

Inserting markers, as shown in Figure 6 - 7, can be handled simply. Select a position on the timeline, choose Marker from the Insert menu and then provide a name for the marker in the field that appears next to the orange flag on the timeline. In this example, the three markers correspond with chapter points for DVD use, each linked to a position where a title screen appears in the video sequence. In a documentary, chapter markers are typically used for major transition points in the work. In a feature film, chapter markers are generally distributed throughout the length of the film and accessible through a scene selection submenu (which can be created in DVD Architect or most other authoring packages). Rather than a title screen being used as a visual divider, the markers in feature films often choose a key visual element that can be readily identified when a viewer is searching through the scene selection options to jump to a certain point in the video.

Figure 6 - 7 **Markers on the Vegas Timeline**

Rendering the Movie

The final stage in the edit process, once you have all the elements combined the way you want them, is to generate a movie. This movie can be targeted for playback through a variety of platforms, including DVD. During the rendering phase, MPEG encoding takes place, if appropriate. When producing a DVD, the goal is to convert and encode all assets to the appropriate format for delivery on DVD. Vegas can accommodate a wide range of source material to accomplish this. Other NLE applications typically have similar capabilities. The File menu of Vegas offers a Render As option. When you select this option, Vegas display the Render As dialog box, as shown in Figure 6 - 8.

Figure 6 - 8 **Render As Dialog Box**

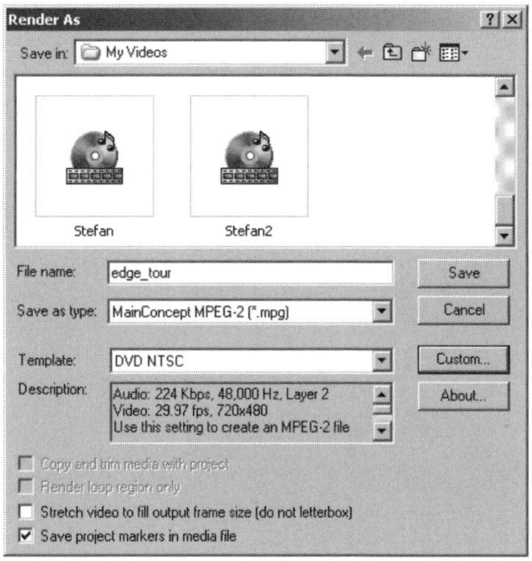

Vegas contains a number of templates that specify what forms of compression should be applied and how the rendering operation is to be handled. These templates can be used to ensure that the correct settings that apply to DVD creation are applied to the audio and video compression. From a single source file, Vegas lets you create any number of rendered files, so beyond just producing the files for an NTSC or PAL DVD, you can generate media in several different streaming formats, such as QuickTime, RealMedia, or Windows Media Video. The output format is directly selected by the Save as type option in the dialog box, which provides access to those templates that apply to the type of output file chosen.

If the available choices are not suitable for a project on which you are working, Vegas includes a Custom button in this dialog box, which lets you specify in detail how to handle the rendering operation. As can be seen in Figure 6 - 9, these options include the ability to control every aspect of the MPEG-2 compression, selecting the bit rates, number of I-frames and B-frames in a Group of Pictures, and defining the field order. These types of options can wreak havoc on your DVD project if you make changes without understanding the implications, but for a professional trying to maximize the quality of the content on DVD, these options can be extremely useful. Since your video source files are not altered by the rendering operation, you can experiment with the different audio and video compression techniques to achieve the best balance between quality and disc space used.

Figure 6 - 9 **Custom Rendering Options**

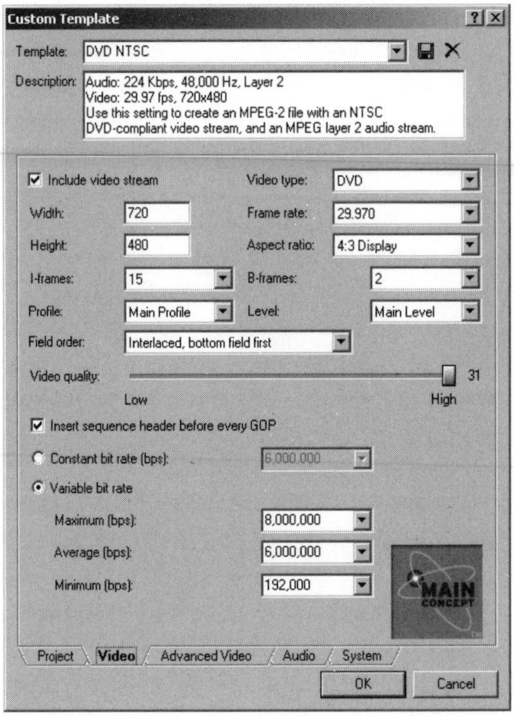

Synergy Between Video Editors and Authoring Tools

It's no coincidence that many software producers offer complementary applications that provide a useful bridge between the editing tasks (including such features as embedding chapter markers) and the DVD authoring tasks. Vegas and DVD Architect include many such links, as do Apple Final Cut Pro and DVD Studio Pro. There are certain tasks that go much smoother if paired applications work together, such as ensuring that an MPEG-2 format file can be directly imported into a DVD project without requiring further compression or transcoding. Although it is not essential to DVD project creation, having complementary applications, or a single application that provides both editing and authoring, can save you steps and help avoid potential incompatibility issues.

7

Encoding Video: Performing Quality Compression

Video compression makes DVDs possible. Without compression, the sheer volume of data required to convey the pixels in a 720x480 frame at 30 times a second for more than an hour would overwhelm both the storage medium and the presentation device, whether a DVD player or computer system. Video compression relies on the fact that much of the information in a sequence of frames is unchanging. Data that doesn't change from frame to frame can be expressed in a shorthand format—for example, if a third of a frame consists of blue sky, that information can be represented in a few hundred bytes rather than the hundreds of thousands of bits it would take to convey the same details pixel by pixel. The technique is somewhat similar to the way that PostScript can compactly represent the information to be printed on a page without having to specify the precise position and color of every single bit composing that page. Video information is encoded (compressed) for storage on disc and then decoded (decompressed) for display on the selected output device. The process requires a high degree of processing power for both the encoding and the decoding.

The shorthand used on DVD to convey video information is MPEG-2. MPEG-2, a compression standard created by the Motion Picture Experts Group, delivers high quality video content at high bit rates ranging from 2 Megabits per second to 9 Megabits per second. MPEG-2 is used on both NTSC DVDs (at 720x480 frame sizes) and PAL DVDs (at 720x576 frame sizes).

This chapter discusses video compression tools and techniques and suggests the best ways to gain optimal quality for video delivered on optical disc.

Compression and Codecs

Audio and video compression are similar in that both rely on codecs. The term *codec* is short for compression/decompression. Codecs use mathematical formulas to reduce or eliminate extraneous information in an image or a sound so that it can be compactly represented for storage. Many codecs focus on specific applications, such as delivering streaming video on the Web (RealMedia) or compressing music for Internet distribution and portable player applications (MP3). The list of available codecs is long and growing, although at any given time earlier codecs are becoming obsolete as newer codecs are gaining increased popularity.

If you are producing DVD-ROM projects, your interest in codecs might embrace the entire spectrum of available compression techniques, since you have the potential of putting any kind of compressed content on a disc. There are multiple possibilities in this area. You might produce a DVD containing several hours of MP3 audio music for playback. A single DVD could contain dozens of short QuickTime movies providing training in tennis or yoga, teaching someone how to construct furniture (for an example, see *Following One's Passion* on page 186), or taking someone on a tour through the back streets of St. Ives.

If you are making a DVD-Video disc, which is a title designed primarily for playback on a DVD player, the video content must be in MPEG-2 or MPEG-1 format. Because MPEG-1 is a lower quality compression format, most DVD titles use MPEG-2 video.

If you are working with source video imported into an NLE application from a digital video camcorder, the video content (usually in AVI or QuickTime format) must be encoded before DVD authoring can be completed. Alternatively, you can use properly encoded MPEG-2 files obtained from other sources for inclusion in your DVD, but some care has to be taken to ensure that the software does not attempt to perform additional compression on the files (unless, for some reason, you want to compress further).

The increased performance of both Macintosh and Windows computers makes it possible to perform MPEG-2 encoding through software, though this can still take several hours to complete, depending on the capabilities of a given machine.

Standalone MPEG-2 encoders can also be purchased. These provide a greater deal of precision specifying the compression properties to satisfy unique requirements. A third option—usually a more expensive option—is to use dedicated hardware encoders (either computer systems designed exclusively for encoding or circuit boards that can be installed in a system to perform this function). High-end hardware encoders, such as the Sony Vizaro, cost in the range of $18,000. These kinds of encoders are usually employed in production environments where the volume of encoding performed on a daily basis dictates a streamlined approach to the process.

For many purposes, the software encoders included with DVD authoring applications generate very good quality encoded MPEG-2. Two outstanding examples are the encoder included in Sony Vegas + DVD that was developed by MainConcept and encoder in DVD Studio Pro, which was produced by Apple. Since these are the typical encoders that will be used by many of the readers of this book, this chapter focuses on their use.

MPEG-2 File Structure

Depending on the capabilities of the compression tool that you use to encode MPEG-2 video, you have control over a number of aspects of the encoding, many of which can dramatically affect the quality of the results. The trade-offs between file size and video quality relate to properties such as the selected bit rate for the video, the degree of compression, the handling of Group of Pictures (GOP) settings, and similar factors. To get a sense of what is involved in the MPEG-2 compression, it helps to understand the basic elements within the file.

Frame Types

MPEG-2 encoding works on groups of frames, rather than individual frames. This technique affords a mechanism by which the redundant information can be identified and related to a frame that contains full information. The three frame types are:

- **I-frames**: I-frames serve as keyframes, providing a complete image of the frame contents at a given point in time. No references to preceding or successive frames appear in an I-frame.
- **P-frames**: P-frames use predictive algorithms to determine the frame content based on the nearest I-frame or another P-frame. P-

frames, sometimes referred to as reference frames, require much less storage space than I-frames. They can be referenced by both surrounding P-frames and B-frames.

- **B-frames**: B-frames interpolate frame contents by referencing I-frames and P-frames around them. This frame type provides the most compact storage method. Highly efficient B-frames are what make MPEG-2 so good at reducing the size of video files.

The following figure shows a collection of frames as they might appear in a GOP. The pattern illustrated is called IBBP.

Figure 7 - 1 **Typical Frame Progression in a GOP**

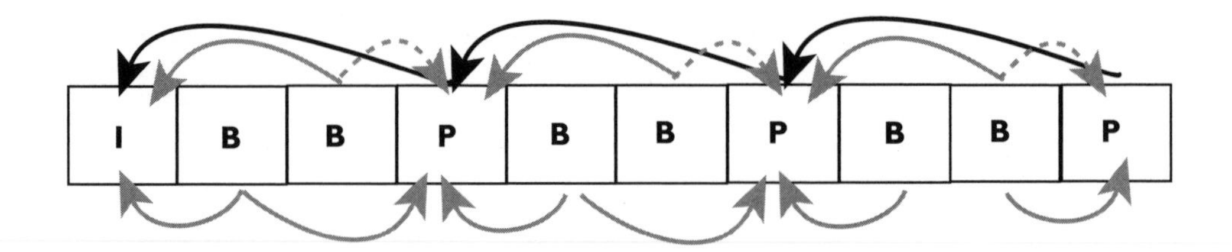

GOP Settings

The Group of Pictures represents the pattern of I-frames, P-frames, and B-frames to be used in the MPEG-2 encoding. This pattern can be modified to produce varying levels of video quality and differing degrees of compression. In some cases, a simpler pattern consisting of just I-frames and P-frames is used.

A GOP can extend for a maximum of 18 frames for NTSC or 15 frames for PAL. Generally, the longer the length of the GOP, the more efficient the encoding operation (since longer GOP lengths have a lower percentage of I-frames). Longer GOP lengths, however, don't work as effectively on certain types of video content, such as very quick camera movements or quick transitions. Certain encoders deal with quick changes in content by automatically inserting I-frames when these changes are detected. In some cases, this kind of operation can be performed manually by someone who is overseeing the encoding operation. The shortest GOP possible consists of a single I-frame and a single P-frame. Such a pattern typically results in larger MPEG-2 files that accurately depict rapidly changing video content.

GOPs can be either open or closed. An open GOP includes B-frames that can examine the I-frame or P-frame from an adjacent GOP. In comparison, the B-frames in a closed GOP are limited to looking at surrounding frames from the same GOP. Although open GOPs function more efficiently at compressing video, they cannot be used if your title includes multiple angles (different camera views). Regardless of what pattern is selected, a GOP cannot end in a B-frame.

Simpler compression tools may not give you the opportunity to change any of the GOP settings. However, if you do have the ability to adjust GOP settings, you can use the characteristics described in this section to select a pattern, length, and an open or closed mode of operation.

Managing Bit Rates

Encoders vary in the amount of control that they provide over the encoding process. Minimally, a developer can usually select the method used for determining the bit rate.

The more sophisticated software encoder applications typically offer three options for handling video encoding:

Constant Bit Rate/One Pass: This method is the simplest and the most effective at predicting the final file size of the encoded video, since a single bit rate is applied throughout, regardless of the complexity of the video content being compressed. Difficult segments with motion and contrast receive the same bit rate as motionless segments with little change in the video image from frame to frame. When selecting a bit rate for CBR compression, a developer should be sure to choose a rate high enough to support the most difficult sections of the content.

Variable Bit Rate/One Pass: During variable bit rate compression, the encoder works from two values that you supply: a basic bit rate and a higher bit rate that applies to segments with greater motion. During the single encoding pass, the encoder detects sections with more motion and bumps the bit rate up accordingly. Otherwise, the basic bit rate is applied to the encoding. The quality of this method is limited to the ability of the encoder to detect and adapt to the motion segments, as well as the limitation of the maximum bit rate selected. Better results are typically achieved using the two-pass VBR technique. Another difficulty with this method is that the compressed video may exceed the available storage space on the DVD.

Variable Bit Rate/Two Pass: The two-pass VBR method provides the highest quality compression. This method also requires selection of a basic bit rate and a maximum bit rate. However, compression is improved by an initial pass that is performed by the encoder to sample the motion content and determine the optimal compression level to use at each point in the video. When the second pass takes place, the encoder draws on this information to both set compression levels, as well as ensure that the final encoded file size stays within a predictable limit so that it will fit on the disc. The extra quality provided by the two-pass VBR method comes at a price—this type of encoding takes twice as long as the single-pass VBR technique.

A current generation DVD player can support a maximum bit rate of 10.08 Megabits per second, which includes all of the compressed video, audio, subtitles, and so on. The maximum video bit rate is limited to 9.8 Megabits per second. Older players may not be as capable of playing back content with sustained high bit rates. Content encoded at high bit rates also occupies more space on the disc and it may not be practical for presentations that include lengthy segments of video (in excess of 100 minutes).

In the Hollywood feature films on DVD, the film itself is sometimes compressed at a high bit rate setting and the extras (interviews and trailers and so on) are compressed at lower bit rate levels. The other option when the content threatens to exceed the capacity of a 4.7GB single-layer DVD-5 disc is to go to a multiple-layer disc. This provides more storage space in exchange for a higher manufacturing cost. For more information about the storage capacities of different DVD formats, refer to Chapter 17, *DVD Fundamentals*.

In real-world conditions, the maximum video bit rate that is generally employed is about 8 Mbps. Bit rates used for MPEG-1 content are much lower with a maximum of about 1.8 Mbps and a typical setting of 1.15 Mbps.

Different DVD authoring applications use different techniques to ensure that the content will fit on the selected media. Making this determination becomes a crucial part of the authoring process—if your video material won't fit on the disc, you can't create the completed title.

Multiplexing Audio with MPEG-2 Video

As a part of the process to prepare files for DVD content, the audio streams and video content are encoded and multiplexed together. This multiplexed combination of audio and video becomes the storage format that is used on the DVD disc. Keep in mind that since the audio and video share space on the disc, if you use uncompressed audio formats, you reduce the amount of storage space available for the video. Uncompressed formats such as WAV, PCM, and AIFF audio take up the most disc space. Uncompressed audio formats, however, provide improved fidelity and audio quality, so for some applications they may be appropriate.

More flexibility is available for the audio content on a DVD than for the video content. The DVD specification supports these audio formats:

- **MPEG-1 Layer 2 Audio**: This audio format has general compatibility with both PAL and NTSC players, but some NTSC players cannot play it back.
- **Dolby Digital audio**: The Dolby Digital audio format (also sometimes called AC-3) delivers 5.1 surround sound, which consists of five primary "surround" audio channels and a central channel sometimes reserved for a subwoofer. Dolby Digital content enjoys universal compatibility with PAL and NTSC DVD players and the compression helps keep the audio file sizes down. This is the recommended audio format to use for most applications.
- **WAV audio**: WAV audio is the dominant audio format on the Microsoft Windows platform, providing uncompressed two-channel content. This format is closely related to PCM.
- **AIFF audio**: The Audio Interchange File Format dominants on Macintosh computers, where it provides two-channel audio content at several sample sizes and rates. This format is also closely related to PCM.
- **PCM audio**: Pulse Code Modulation audio is the native format that applies to CD delivery of audio content. Unlike the CD, however, the DVD specification supports uncompressed PCM at higher sample rates and sizes. The basic CD rate is 16-bit samples at 44.1KHz. The DVD specification includes sample sizes of either 16 bits or 24 bits and sample rates of 48KHz or 96KHz. Universal compatibility with players is achieved with this form of audio at a high degree of fidelity. Since the audio is uncompressed, however, it consumes greater amounts of disc space.

A DVD project can include a mix of audio formats, although formats must be consistent for specific elements, such as the audio corresponding with the DVD menus.

The bit rates within those elements in which the audio must stay the same can encompass a range from 64 Kilobits per second to 4608 Kilobits per second (depending on the type of audio—not all types support all rates). Within these bit rates, other settings must not be mixed within certain elements. The sample rate should be either 48KHz or 96KHz. The sample size should be 16 bits or 24 bits. The number of channels, if using Dolby Digital audio, should remain consistent. The audio type also should not be mixed. A single type (AIFF, WAV, PCM, Dolby Digital, or MPEG-1 Layer 2) should be applied consistently to the appropriate elements.

Audio that is not available in one of the formats supported by the DVD specification can usually be transcoded into a supported format. This transcoding can often be accomplished within an encoder (for example, the QuickTime MPEG Encoder used by Apple in DVD Studio Pro can convert MP3 audio files into uncompressed AIFF for use in a DVD project). You can also convert audio content using any number of audio editing applications. Bias PEAK on the Macintosh supports a wide range of formats and conversions. On the Windows side, Sony Sound Forge provides flexibility in converting audio from one format to another. If, however, you convert from a compressed format (such as MP3) to an uncompressed format (such as WAV or AIFF), there is some loss of quality when compared to the original source material.

Guidelines for Achieving Optimal MPEG-2 Encoding

The character and quality of the MPEG-2 encoding in a DVD-Video title makes an enormous difference in the perceived quality of the title. Poorly compressed or over-compressed video exhibits many annoying artifacts and visual disturbances. To get optimal quality video during compression, follow these guidelines:

- **Work from the highest quality source files**: The quality of the source content for MPEG-2 encoding limits the ultimate quality of the output. Start from uncompressed digital content whenever possible. DV formats offer the next best choice, followed by analog Component content. If you are limited to consumer-level analog video options, S-Video provides significantly better results than Composite varieties of analog (which should be avoided for MPEG-2 compression).

- **Avoid noisy video content whenever possible**: Video that contains significant amounts of noise adds to the difficult of encoding, since the encoder attempts to replicate each element of the noise (instead of being able to identify and store large regions of unchanging image). Starting from content with noisy video generally produces larger encoded files containing poorer quality video.
- **Use two-pass VBR encoding**: If two-pass VBR encoding is supported by your encoder, this option often provides the highest quality results. The bit-rate can be adjusted on the fly to compensate for changing areas of video, which are detected during an initial pass through the file. Scenes containing motion are generally handled in a much cleaner manner using two-pass VBR techniques, but allow extra time for this process since it takes twice as long as single-pass VBR.
- **Start with a video source that complements the chroma format**: The chroma format, or color space, used in the DVD specification is 4:2:0. Video content from sources that use other chroma formats (such as the NTSC DV format with a 4:1:1 color space) require conversion, resulting in some color loss from the reduced sample size. The PAL DV format, which uses the 4:2:0 color space works well for DVD, as well as the higher end professional DV formats, such as DVCPRO 50, which uses 4:2:2.
- **For difficult source video, use a software encoder with extended encoding times**: Even high-end hardware encoders have limitations in terms of the quality of difficult video content that they can produce, largely because they are typically designed for high-volume applications in industrial settings. Software encoders, in comparison, can be slowed down so that they can take an extended period to examine and optimize the encoding process. For difficult video, where content changes frequently or contrasts and color shifts can be dramatic, a software encoder can generally produce better results. For an example of this, refer to *Getting the Best Video Compression Results* on page 137.
- **Examine your results critically**: For important projects where the quality of the encoding is critical, perform some tests. You can burn interim copies of a DVD using encoded video that has been generated at different settings. Examine the results side by side—the differences in quality caused by varying the encoding settings are often strikingly obvious. If you feel that you've become too close to the project to make a reasonable judgement, bring in a

neutral observer—someone who hasn't ever seen the work—to assess the relative quality of the video. A neutral observer can often provide important insights into the appearance of the images that you might miss.

Summary

All of the care and attention that you pay to your audio and video content during the editing process can be lost by making poor choices during encoding. To get maximum results for your efforts, you need to remain keenly aware of the trade-offs between file size and quality during each stage of the DVD authoring process. Automated compression settings can work well for many types of general-purpose content, but if you understand the nature of compression, you can often modify compression properties to ensure excellence in the quality of the audio and video on disc.

Beyond the basic rules and guidelines, there is no substitute for experience and experimentation in this area. Video encoding may seem like medieval alchemy until you become familiar with the options and settings available for encoding video. The more you experiment with different types of video content and honestly assess the results of the encoding, the closer you will be to producing top-notch compressed video for presentation on DVD.

8

Authoring a DVD Disc: Creating a Title

Perhaps the most satisfying moment in the course of making a DVD is when you begin authoring, taking the full collection of audio and video files that you have produced over weeks or months, and assembling this content into a DVD disc playable on hundred of millions of players.

Authoring tools range from the ridiculously simple to bewilderingly complex, but most current-generation tools can be mastered without enormous difficulty and used reliably to produce DVDs that can be used directly for distribution, for duplication, or for replication.

That being said, there are a fair number of choices that need to be made for even a relatively simple DVD. Should the video content begin playing immediately or should the viewer be directed to a main menu? Will the disc contain features for computer playback that won't be available with set top players? Will there be a number of submenus providing access to extras and special features, or will the navigation consist of a few buttons on a single menu?

To some degree, your options will be limited by your choice of authoring tools. This chapter provides examples using two very capable tools: Sony DVD Architect and Apple DVD Studio Pro. Both of these tools offer a very full range of options, with DVD Studio Pro providing near complete mastery over the authoring process. For initial projects, you may want to work with simpler authoring tools, but when creating DVDs for professional or commercial distribution, the higher end tools include key features that might be required by your project, such as the capability of specifying an encryption scheme for copy protection or other special features that aren't available on the entry-level authoring tools.

DVD Creation with DVD Architect

The Edgewood Studios tour that appears on the DVD bundled with this book provides a good example of a project that can be rapidly converted to a DVD title. Constructed in Vegas, the project was then rendered using the DVD NTSC template in Vegas. The resulting compressed MPEG-2 files, created using the MainConcept MPEG-2 encoder, can be directly imported into DVD Architect without further encoding or transcoding. Whenever possible, avoid performing multiple compression passes on the audio or video files—quality suffers during each compression operation. DVD Architect opens to the interface shown in Figure 8 - 1.

Figure 8 - 1 **DVD Architect Application Window**

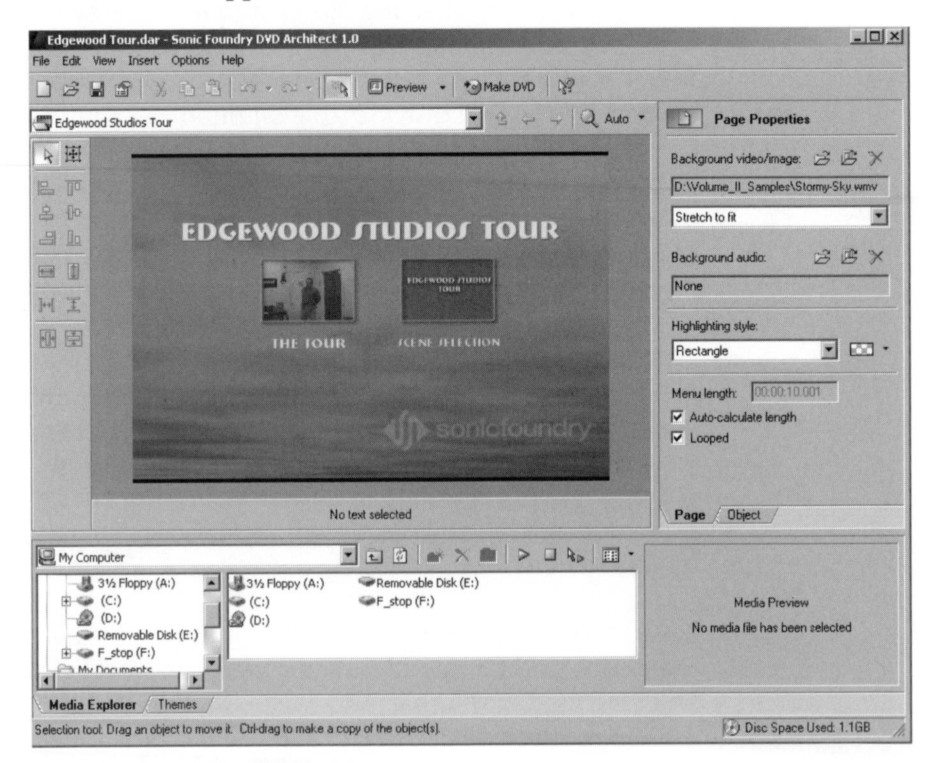

DVD Architect works most efficiently when the files to be used on the DVD are already in one of the native DVD file formats. Other file types can be used, but in most cases, DVD Architect must convert them to a compatible format, which increases the time required to create the DVD and sometimes degrades the quality of the content.

Front End Design

The background video or image can be selected either using the Media Explorer or through the drop-down list box under Page Properties. In this example, the background is selected from the Sony Vision Series (see *Vision Series Themes* on page 90).

Optionally, you can select an audio backdrop to play while the menu page is being viewed. Some themes contain their own audio, in which case you can quickly enable the background audio by clicking the From Themes button. You can also select your own audio file or choose not to have any audio play during the menu selection.

The Insert menu lets you create a submenu linked to the main menu page or to add buttons that link directly to media content (such as audio or video files). In the example shown in Figure 8 - 1, the button labeled The Tour leads directly to the video and the button labeled Scene Selection opens up a submenu that contains chapter links to navigate within the video segment (as shown in Figure 8 - 3).

Project Types

The type of project selected in DVD Architect determines what options appear in the initial application window. The simplest project to create is a single movie, which can start playing immediately upon insertion in a DVD player. To do this, you open a new project and select the Single Movie option and DVD Architect prompts you for the video file to use. You can select start and end points for the video clip you choose, but the Single Movie option doesn't allow the creation of menus.

Music Compilations and Picture Slideshows are two other project options that don't require menus, and these both can be created rapidly simply by selecting the files to play in the correct sequence.

For many DVD titles, you will probably use the Menu Based option, which allows you to flexibly create a main menu and a number of sub-menus. Users can freely navigate to the content they want to play and return to the main menu or any of the other menus to move through video segments. Chapter markers can be used to provide convenient entry points into long video works. DVD Architect can also recognize markers inserted while editing in Vegas, so you can position all of your chapter entry points before even opening DVD Architect.

Commercial DVD titles typically open to sample video clips accompanied by music, eventually leading to a menu. Menus often include small thumbnails within buttons, containing a few seconds of the video to which the button leads. The backgrounds of a menu are often animated as well, and audio (usually looped) plays during the opening sequences in either the main menu or any of the submenus. DVD Architect supports these types of design decisions.

Vision Series Themes

To provide a professional finish to a DVD project, you can use themes, which generally contain a video background, button designs, menu lay-out templates, and sometimes audio backdrops. Sony offers prebuilt themes that can be used with DVD Architect projects in the Vision Series (shown in Figure 8 - 2). Samples from two of the Vision Series volumes (formerly from Sonic Foundry) are included with DVD Architect and you can purchase additional themes to match project types.

Figure 8 - 2 **Vision Series Themes**

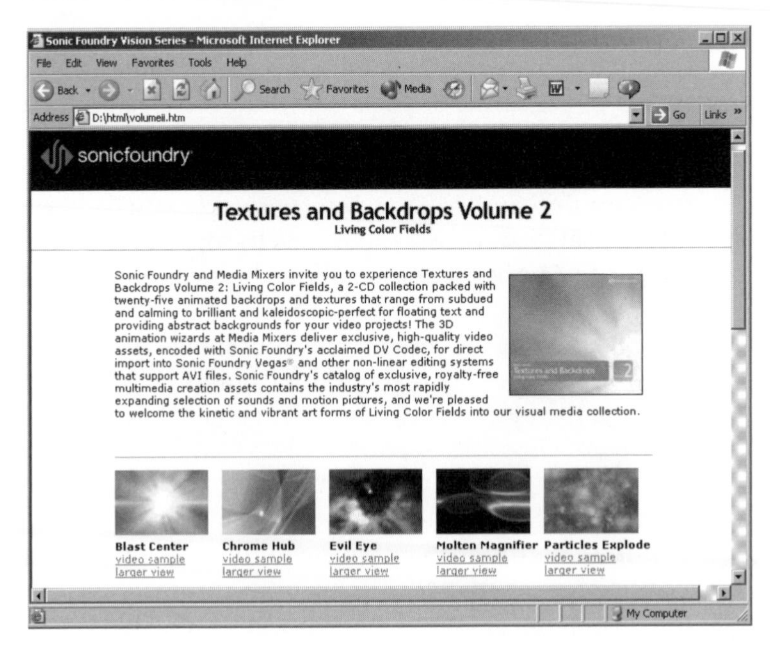

Creating Scene Selections

A scene selection menu, as often used with commercial feature films DVDs, provides chapter entry points to jump to individual parts of a movie. DVD Architect makes it easy to construct a scene selection menu, providing a command on the Insert menu that guides you to a page based on the number of selection options you want to provide, as shown in Figure 8 - 3.

Buttons can be created using a range of graphic frame types and then link information assigned to each button to determine the entry point into the video. If the video does not already have markers embedded in it, DVD Architect lets you create chapter markers on a timeline that corresponds with the video actively being displayed, as shown in Figure 8 - 4.

Figure 8 - 3 **Scene Selection Design**

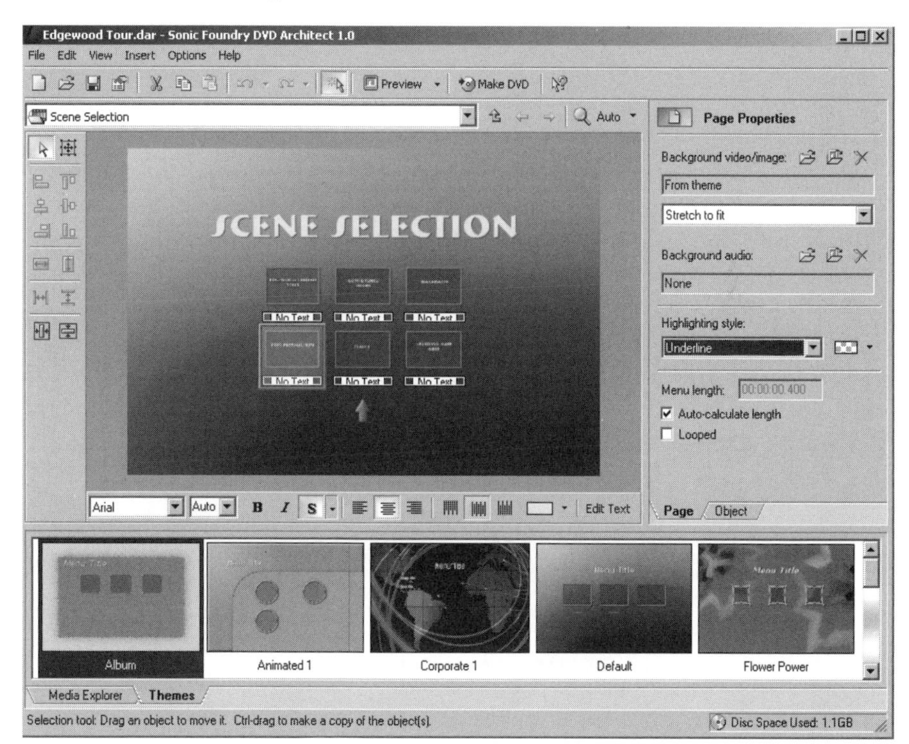

The Media Properties panel and the timeline that appears below the preview window let you create markers that correspond with a particular segment of video. Once a marker is in place, it can be referenced with a link from any button on the scene selection page.

DVD Architect lets you freely create, modify, and delete markers as necessary to construct an appropriate navigation structure to present any type of video content.

Figure 8 - 4 **Chapter Markers**

Previewing a Title

During the DVD title creation, you can check the progress of your work at any time using the Preview command, which appears both as a menu option and button on a number of screens. The preview feature brings up a display window that starts at the beginning of the video, provides live menu access to individual sections, and responds to typical remote control commands (using the simulated controller at the right side of the page).

A video safe and title safe region can be displayed as dotted rectangular boxes to show you how the content will look displayed on a typical television monitor. Standard monitors use overscan, which cuts off the edges of the image. This video and title safe regions are shown in Figure 8 - 5.

During the design process, you will probably skip back and forth between menus and previewing several times before you get all of the content organized the way you want it. Because DVD Architect doesn't do any actual rendering when the preview functions take place, the operation is fast and seamless, letting you quickly get a sense of how the DVD will appear when finished. You can try out multiple layouts and experiment with the organization of the menu options until the timing and graphic design suits your preferences.

Figure 8 - 5 **Preview Window**

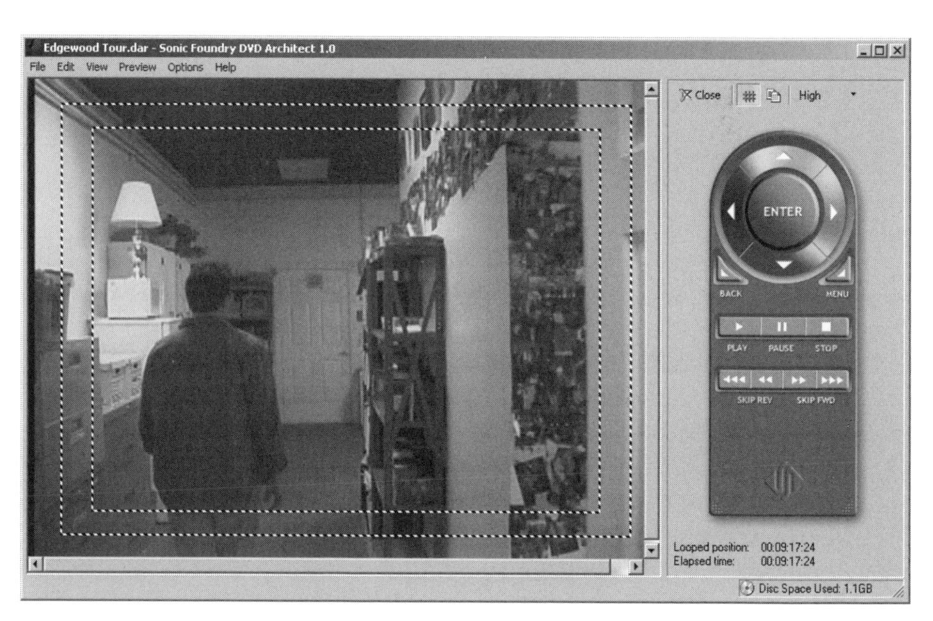

Make DVD

The Make DVD button, displayed on the DVD Architect button bar near the top of the window, launches a dialog box that lets you prepare the files for disc burning and then burn the completed disc. Once you have completed the Prepare DVD tasks, you can use the Burn command in the future and DVD Architect will access the previously prepared project files to speed up the disc recording operation. The file preparation often takes longer than the actual disc burning. Any time you make a change to any aspect of the project, even if it is just slightly moving a button on a menu, DVD Architect has to repeat the Prepare DVD operation before it can burn a disc that includes your modifications.

On any project, run one or more check discs and try them in many different players (including both set top players and computer DVD-ROM drives) to test both the compatibility, as well as the overall functioning of the title. Be sure to run through each of the menu options and navigation buttons to make sure they respond as anticipated. Even a simple DVD menu usually includes enough choices that it is easy to create an inappropriate link or fail to connect a button to the correct chapter marker. Particularly if you are sending a DVD title out for replication, spend a good percentage of your time in final testing and enlist help from anyone willing to spend the time checking and rechecking the navigation features. This work, while tedious, helps avoid the grief of making many copies of a flawed disc.

DVD Creation with DVD Studio Pro

Apple fostered a creative revolution with the release of DVD Studio Pro. The esoteric art of producing a DVD title was suddenly accessible, brought to a level where the skills can be mastered on the desktop by anyone with enough interest and a compatible system. In the intervening generations, new authoring applications have been produced—both for Macs and PCs—but many professionals continue to stick by DVD Studio Pro for its combination of intuitive operation and depth of features. Version 2.0, released in late summer 2003, incorporates a number of capabilities from Spruce Technologies, a company that Apple acquired some months back.

DVD Studio Pro includes a number of features that are significant for someone seriously involved in DVD production. For example, support for the DVD-R Cutting Master Format (CMF) is included in version 2, making it possible to use authoring media to submit a disc to a replicator, enabling copy protection features. Added dual layer support has been incorporated for DVD authoring, so that an author can select a Parallel Track Path (PTP), where both layers play in a single direction, or Opposite Track Path (OTP), where the second layer plays in the opposite direction of the first. DVD Studio Pro 2 can also read in files from DLT, making it easier to copy and transfer disc images to another media or to start a new project from DLT files. These and other advances have made DVD Studio Pro a genuinely professional tool for DVD authors.

Studio Pro also provides a very flexible interface, letting the developer organize one to four quadrants that contain selected tabs for performing certain operations. You can design a configuration that suits your work preferences and put those tools that you use frequently within easy access.

This section examines the construction of the DVD disc included with this book.

Source Materials for Bundled DVD

The source materials that appear on the bundled DVD came from a variety of sources and an equal variety of media, from miniDV cassette to recordable DVDs. The Import button, shown in Figure 8 - 6, lets you select the content that you want to include on a DVD title. This can include MPEG-2 or MPEG-1 video content, which Studio Pro accepts in demultiplexed form (with the audio and video existing as elementary streams). You can bring in video content from other formats as well, such as AVI or QuickTime, if it is a format that can be transcoded or encoded by the QuickTime MPEG Encoder. Ultimately, all video content must be DVD compliant before Studio Pro can include it in a track.

Figure 8 - 6 **Assets window**

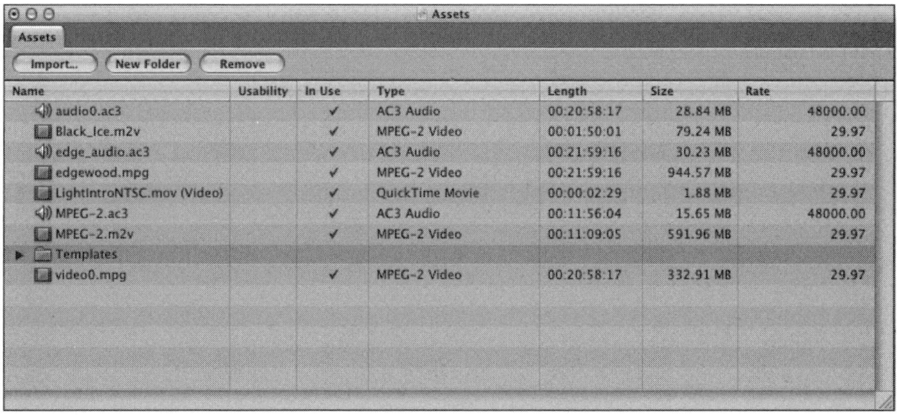

Assets

Import... | New Folder | Remove

Name	Usability	In Use	Type	Length	Size	Rate
audio0.ac3		✓	AC3 Audio	00:20:58:17	28.84 MB	48000.00
Black_Ice.m2v		✓	MPEG-2 Video	00:01:50:01	79.24 MB	29.97
edge_audio.ac3		✓	AC3 Audio	00:21:59:16	30.23 MB	48000.00
edgewood.mpg		✓	MPEG-2 Video	00:21:59:16	944.57 MB	29.97
LightIntroNTSC.mov (Video)		✓	QuickTime Movie	00:00:02:29	1.88 MB	29.97
MPEG-2.ac3		✓	AC3 Audio	00:11:56:04	15.65 MB	48000.00
MPEG-2.m2v		✓	MPEG-2 Video	00:11:09:05	591.96 MB	29.97
▶ Templates						
video0.mpg		✓	MPEG-2 Video	00:20:58:17	332.91 MB	29.97

MPEG-2 files that are brought in from other sources, such as other encoders or non-linear editors that export MPEG-2 as a final step, display a red button in the usability column, which lasts until Studio Pro either determines the file is DVD compatible or converts it for use in the project. Additional encoding and transcoding may take place if a file is too large for an existing single- or dual-layer title—avoid this extra step, which can introduce artifacts, by monitoring file sizes during DVD creation and by importing files encoded to the degree of compression that you prefer to use in your project.

Other assets, such as graphics file and audio content, can be imported for use through the Assets window controls, as well. Studio Pro will convert these, as required, to a suitable DVD compatible format. As shown in the figure, the Templates folder appears as an asset, giving you access to standard DVD ready templates or intro videos that can be used to accelerate the title production and present a unified interface.

Applying a Template

DVD Studio Pro includes a number of templates, shown in the palette in Figure 8 - 7, which control the appearance of menus, buttons, and other DVD features. If you have been working with iDVD, Studio Pro will also import any templates you have been using or that you have created under iDVD. You can also create custom templates to use on a regular basis or for a single title.

Figure 8 - 7 **Templates presented in palette**

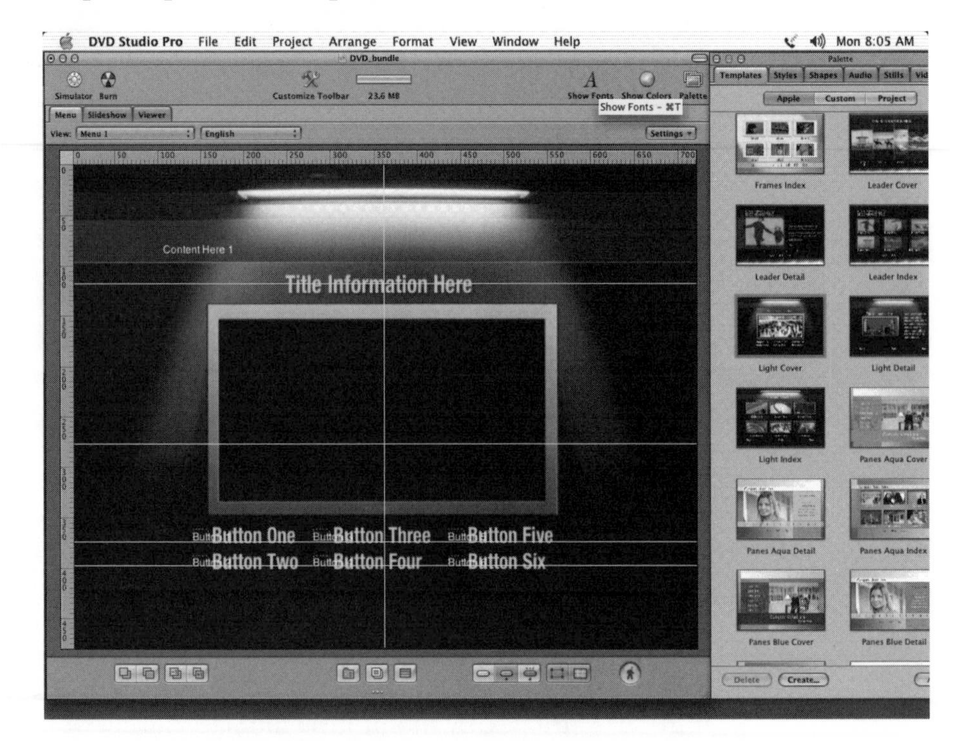

Some templates include a variety of advanced features, such as the drop zone area on the template that was used for the bundled DVD (the Light Cover template). The drop zone lets you drag a video asset into the designated area for playback in the simulated screen in the center of the menu. The Light Cover and Light Detail templates, for example, represent two related designs that provide the opening menu and underlying menus.

The standard button placements and title areas shown within the template display, which are known as objects, can be easily modified by opening the Inspectors pane and performing any necessary changes. Inspectors exist for all of the fundamental components of the DVD title, including the tracks (where video assets and audio content are assembled), the menus and submenus included in the project, and the connections between individual elements (which determine the navigation elements guiding movement through the DVD contents).

Once the relationship between the DVD components and the inspector controls becomes evident, developers can very rapidly make modifications to the DVD contents, setting up buttons to link to video segments, choosing introductory videos to display when a disc is inserted, or choosing a portion of a longer video to include by specifying start and end points in the tracks inspector. Although rudimentary editing and selection can be performed through DVD Studio Pro, the application does not substitute for the functions in a non-linear editor. These extra capabilities simply give you some additional flexibility when dealing with video assets; for example, if you wanted to identify a 20-second clip to use as part of the introductory section of the title.

Designing a Menu

Menus provide structure to a DVD title. The person viewing the DVD uses buttons on different menus to navigate through disc contents and access tracks.The design of each menu includes a background, which can consist of a still image or a movie that occupies the full frame. Studio Pro supports two menu design techniques. The standard method provides the most flexibility, supporting full-motion backgrounds and overlays to buttons to indicate button states. The layered method offer less flexibility, requiring a still image for the background, but providing more elaborate button activity states.

Templates vastly simplify the process of making a menu, providing all the standard components needed in a single cohesive design. The developer can connect buttons to assets and specify a few project details to quickly launch a title. Many templates include the ability to support related menus, such as a initial menu that appears following the DVD intro followed by a submenu that might include links to individual chapters within a title. For demonstration, the following menu shows links to four of the video features included on the bundled DVD (without requiring use of a submenu). In practice for more complex titles, you may want to divide your content for better navigation and include additional menus that access chapters, as is typically the case with commercial DVDs for feature films. Alternatively, you can create a simple DVD that launches and plays a movie from start to finish upon insertion in a player, a technique that can be useful for marketing or promotional videos, or for shorter features, such as animated works or short documentaries.

The Menu Editor, contained within the Menu tab, provides precision rulers, language selection, and the detailed setting necessary to control every aspect of the menu design. A number of alignment features and guides help line up text and buttons on the menu, letting you snap an object into a particular position or create a column of precisely aligned objects.

Figure 8 - 8 **Main menu construction**

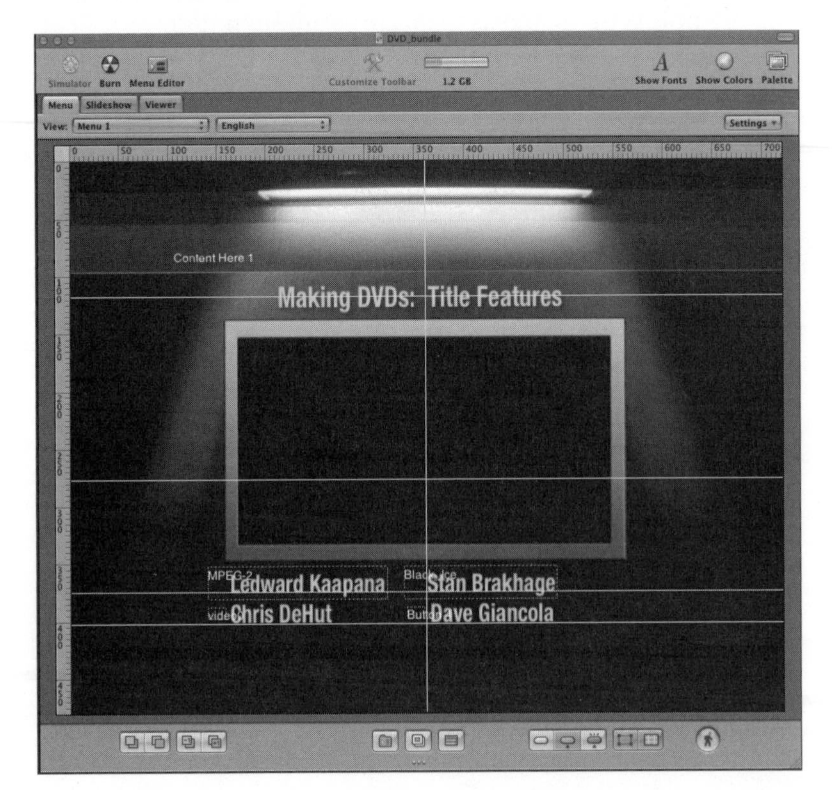

Using Outline View

The outline view of a DVD project list the components and lets you navigate to certain specific elements, such as an individual video track. The outline view, as shown in Figure 8 - 9, offers a broad overview of the DVD development, displaying the menus that you have designed, the tracks designated for the project, slideshows available on the disc, and scripts that have been applied. You also can view the language support that has been defined for the title.

Figure 8 - 9 **Outline view**

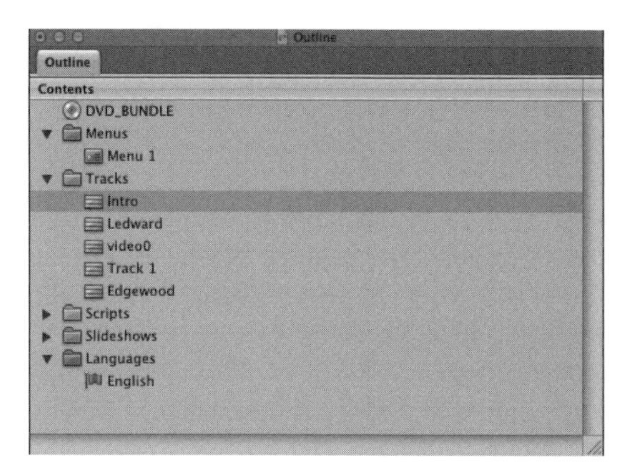

Dragging assets to the Outline tab is a quick way to build the contents of a project. Drag and drop actions support audio, video, and still images. The use of the object brought into the outline depends on where you drag the asset. For example, if you drag a video assets to the disc's name (in this case, DVD_BUNDLE), Studio Pro creates a track in the project and adds the video to stream V1. If you drag a video asset to the Menus folder of the outline, Studio Pro generates a new menu and uses the video as the full-motion background.

Similarly, you can add an audio backdrop to a menu by dragging the audio asset to the appropriate menu. The outline supports a full range of drag-and-drop operations which can be a helpful way to bring together the aggregate design elements and contents of a project.

Double clicking on a track lets you view the contents of the track, as shown in Figure 8 - 10.

Figure 8 - 10 **Track display**

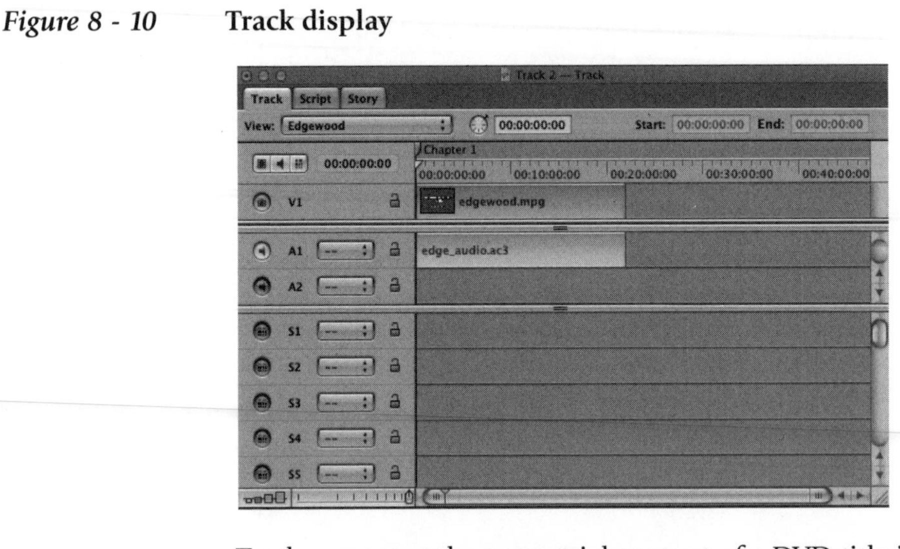

Tracks represent the sequential content of a DVD title in a timeline format. A track can contain a segment of audio, video, or set of subtitles organized to play in a linear manner. A single track can contain a significant amount of content, including 9 video streams (offering angles or alternate video selections), 8 audio streams (providing support for alternate sound tracks, such as different languages), 32 subtitle streams (for overlaying text onscreen for various purposes), 99 chapter markers, and 98 stories (stories offer a method for directing playback along specified paths).

A trim feature lets you perform rudimentary editing, determining what portions of each asset will be included in the segment. Depending on the nature of your project, you may want to create a single track that plays in sequence, or a series of multiple tracks that can either be linked or played individually. For a continuous, glitch-free viewing experience, a single long track is best—DVD players sometimes pause when moving from one track to another.

The Track inspector provides control over individual elements of each track, letting you define an end jump point (connecting the track to another track or a menu), a display mode, closed captioning features, and other options.

The streams shown in the window of the Track tab let you control the relationship of the individual assets, functioning in much the same way as the audio and video tracks in a non-linear editor. You can turn audio or video streams on or off, create composites of individual streams, and perform many other functions that control how the track will be played in the final title. The streams are displayed using a timecode that helps to precisely determine start and stop times for clips and to otherwise control stream properties.

Previewing a Title

As with other authoring applications, Studio Pro relies on pointers to indicate the content contained in a title, so most projects can be previewed very easily without requiring final rendering (which is often a time-consuming process). The Simulator window, shown in Figure 8 - 11, lets you navigate through the menus and test buttons and see the title as viewers will see it in final form. The Simulator window makes a great debugging tool and should be used extensively to check operation of a title's navigation features.

Figure 8 - 11 **Preview display**

If you detect problems in the way that menus operate or in the sequence of tracks displayed, the Connections tab offers a global view of the links between individual objects in your project. You can also modify and adjust these links, providing a central point for controlling many of the individual connections that are created in other tabs, either automatically or through your use of inspectors and other controls. The Connections tab is a good place to ascertain the workability of all the elements of a project before you commit to building and burning it to disc.

The Disc Inspector also provides a useful tool for examining (and modifying, if necessary) any of the DVD options before proceeding to building and burning of the disc. You can open the Disc Inspector by selecting the name of the disc from the Outline tab. Options can be specified, such as what segment will play first when the disc is inserted, menu display options, selection of subtitle streams, Macrovision settings, and so on. These options are generally referred to as the prebuild disc properties and they determine how the elements in your project are combined to comply with the DVD specification.

Burning the Disc

Studio Pro lets you precisely control the burning of the completed title to DVD, or, optionally, you can output the project specifications to DLT in DDP format or CSS format (which includes copy protection options). Whether you choose to build the disc first (which stops the process short of actually writing to a DVD or tape) or select the burn command (which first builds and then burns the DVD), similar decisions must be made to set the build and format configuration.

Developers typically perform a number of incremental builds of a project to test and ensure that each element used in a title is functioning properly. Elements from prior builds remain in their compiled set. This helps reduce the time required for each subsequent build of a title. The build operation rechecks the VIDEO_TS folder and uses any unchanged compiled elements before proceeding to prepare the other elements.

Most typical single-layer projects can be set up to build automatically, whereas dual-layer projects introduce additional complexities that require more oversight. As the formatting is taking place, Studio Pro requests that a second recordable disc or a blank DLT cartridge be inserted before the second layer is written. Also, the determination of the break point in a long sequence often requires manual intervention. Dual-layer projects in general require more attention than single-layer projects. Singe-layer projects can often be built and burned unattended.

Figure 8 - 12 **Building and burning a disc**

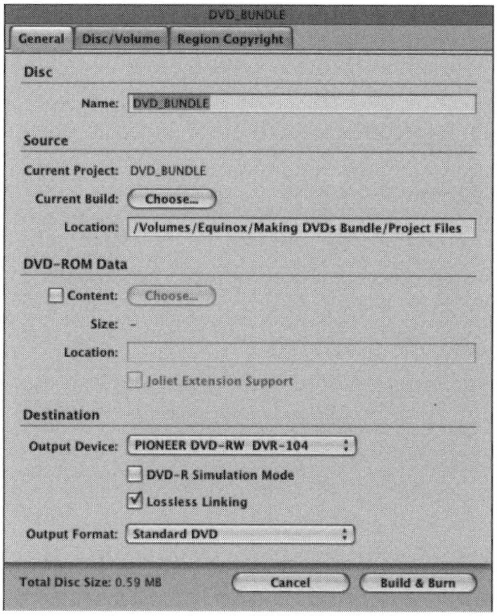

If your project includes DVD-ROM data, checking the Content option lets you select the location of the files that you will include in the disc ROM folder. Files in the ROM folder can be viewed when the disc is played in a computer, but they have no effect on any set-top DVD player options. Developers often use the ROM folder to store materials that are related to the disc contents. For example, Chris DeHut uses the ROM folder when producing his *Woodworking at Home* DVD magazines (discussed in Chapter 14, *Case Study: Producing a DVD Magazine*) to store the plans for woodworking projects in Adobe Acrobat format. You can use the ROM folder for virtually any kind computer files—since the file structure uses UDF architecture, the contents of the ROM folder should be viewable on any computer platform that supports UDF file reading. All recent Mac or Windows computers have this capability.

Studio Pro maintains a running log of the tasks and completion status during the build and burn operation. If you run into any difficulties successfully recording the DVD content to disc or tape, this log will potentially clarify the point at which the problem occurred so that you can make configuration changes to ensure seamless operation.

Extending the Possibilities

This chapter includes only a brief sample of the full capabilities of DVD authoring tools. Simpler authoring packages rarely give full access to the extended range of possibilities, but applications such as Apple DVD Studio Pro and Sony DVD Architect support the majority of available features (with Apple having an edge in this department).

Many resources exist for furthering your knowledge about DVD authoring, ranging from training videos on disc to tutorials produced by the independent software vendors who create many of the authoring packages. The skills and expertise needed to fully master the possibilities of authoring may take years to acquire, but you can become conversant with the most important features of a typical authoring application in less than a week. The more you work with the medium, the more adept you'll become at mastering it.

9

Case Study: Pushing the Medium

The interactive character of DVD and its flexibility in dealing with different forms of digital content have encouraged some DVD producers to stray from the predictable path and attempt more ambitious projects. *1 Giant Leap*, released by Palm Pictures in 2002, explores some of the key themes of civilization in video collages, music, and interviews with a number of the most interesting thinkers of our era. Designed from its conception as a DVD presentation, *1 Giant Leap* incorporates innovative features, such as an electronic jukebox that can cycle through the various chapters included on the disc in different sequences.

The project also highlighted the advantages of digital video cameras and non-linear editing tools. The two filmmakers, Jamie Catto and Duncan Bridgeman, travelled light during their journey across Africa, Southeast Asia, Europe, and North America. The lightweight DV camcorders performed admirably in difficult conditions. Shooting primarily under natural light, the filmmakers shrugged off heat, rain, dust, and insects in their travels. The resulting multimedia title is a composite of world music, insights on such heady topics as Time, God, Sex, and Death from speakers such as Kurt Vonnegut and Tom Robbins, glimpses of different societies from around the world, and an underlying theme that our essential humanity unites us all.

National Geographic Channels International (NGCI) partnered with Palm Pictures to present a documentary special, *The Making of 1 Giant Leap*, which premiered in April 2003.

Origins

Chris Blackwell, the founder of Palm Pictures and one of the guiding spirits behind *1 Giant Leap*, sees DVD technology as liberating to creators. Palm Pictures has developed a catalog of properties, including many DVD releases, showcasing unique and original works from independent filmmakers, animators, and musicians. As an inductee to the Rock and Roll Hall of Fame in March 2001, Chris had an open ear when approached by two young musicians about the project.

"The guys who put together the music," Chris said, "Jamie Catto and Duncan Bridgeman, came into our office in London and pitched us on allowing them to go to different parts of the world to complete a recording, which would be a mixture of different music genres. They played us a demo that was made up of samples and they wanted to go around the world to record the people live. So, I said, 'Well, you know, Palm Pictures is set up as a DVD company. I love the music, and I love what you guys are doing, but I'd like to add to it. You should bring along a couple of people and film everything so we can make it into a DVD.' And, that is exactly how it happened."

The overall theme of the project evolved organically as the music and video was captured and the project team began sorting through the assets. "We let Jamie and Duncan pull the material together," Chris said. "They would come back and show us what they had done. The project was pretty much all done out of England. The thing that took most of the time, as you would imagine, was all the editing—a tremendous amount of time. Eighteen months or more."

Authoring the DVD

David Beal, President of Palm Pictures, oversees many of the technical aspects of DVD projects. He pointed out that the authoring process for *1 Giant Leap* was technically challenging because of the unique elements that were worked into the design.

"The authoring was quite extensive," David said, "especially when we looked at the programmable jukebox function and the explorer function. The bit-budgeting on that disc was pretty much completely to the max. We almost didn't get the last music video on, because we couldn't squeeze the content enough to fit."

Depending on whether the video footage was designed as value-added footage on the extras portion of the disc, or the primary feature footage, the bit rates were scaled to the level of quality desired, with variable bit rates being used when possible. "Obviously, you want to do everything variable—you have to be really careful to maintain the look you want when coming off DV. Most of this was shot on DV with some Betacam."

Authoring and many other parts of the project took place at the Palm Pictures satellite office in England. During the course of development, Suzette Newman, Jumbo Vanrenan, and Adrian Boot were very active in the process. Apple Final Cut Pro was used for the video editing, but the complexity of the jukebox and explorer functions went beyond the capabilities of DVD Studio Pro. Instead, the final stages of the DVD authoring included scripting and hands-on work in DVD Producer.

The Compelling Nature of DV

Much of the footage in the project was recorded on DV format cameras, a technique that is used with many Palm Pictures projects that feature dynamic content and also require unobtrusive video work.

"I attribute the fact that we used DV," David said, "to why a lot of the footage is so compelling. This is something that Chris has always highlighted: when you go into a situation with a camera rig and lights and all this equipment, it changes the situation. A lot of the footage was captured by the team when they were using small handhelds out in the middle of the bush in different places. The equipment is so unobtrusive that people have a tendency to forget about the filming process. You get much more compelling video."

CD or DVD?

During development, the staff at Palm Pictures struggled to resolve an important issue. Should *1 Giant Leap* be first released as a songs-only CD or as a DVD with the full audio-visual experience?

Chris remembers the thinking at the time, "You know, *1 Giant Leap* was the first of its kind—something that was actually conceived as a music-based DVD. We felt then (and now) that the music, the audio side, would probably be the best way to start bringing it some attention. We went backwards and forwards on it, for about a year, trying to decide whether we should release the DVD first, or release the CD first to get some attention for the upcoming DVD. We eventually decided to release the CD first and then release the DVD."

Breaking Down Barriers

Part of the impetus behind Palm Pictures is breaking down the rigid barriers that have stifled many creators, both in filmmaking and the music world. The shift is sweeping and undeniable. David said, "I think the *1 Giant Leap* project is very symbolic of what we have seen as a revolution in filmmaking. It is somewhat equivalent to what the advent of the electric guitar did to music, as well as what the turntable did to music when it became an instrument. I think that all of these new digital creator tools, whether the DV cams or Final Cut Pro or Avid DV Express, are changing the type of people who can make films now. People can dabble and when they start to get something good, it is not like they are on a budget and have to finish it in a month. With this approach, you are going to get some really interesting rock-n-roll filmmaking."

The independent filmmakers may be able to move into the vacuum created as Hollywood has increasingly aimed for the mainstream markets. "I think that Hollywood has had the traditional feature covered for a long time," David said. "The new stuff that takes advantage of the new formats and the new creator tools is more interesting to us."

Palm Pictures is currently working on the *Directors Label*, a series of DVDs that highlights the work of prominent directors. "The first three," David said, "are Spike Jonze, Chris Cunningham, and Michel Gondry—all guys who have really pushed the boundaries with visual work. I think that prior to the DVD format, this type of project wouldn't have really made it. It's not the kind of thing where you just put a whole batch of videos on VHS. The approach is very non-linear with short films, commercials, music videos, and commentary by the artists and directors. It is incredibly entertaining, but you also learn a lot about the process by listening to what these people went through. DVD is the format that has enabled these types of projects to come to fruition."

Chris said, "We're really excited about the *Directors Label*. I think they are a sort of Exhibit A—the kind of thing that I believe will really happen with DVD. Basically, you are examining the creative force behind a whole lot of videos that you have seen. You are letting people in to look at the body of work of this person. So, there is a medium through which you can actually promote a director, which never really existed before."

Counting on DVD

Despite the extremely slow ramp-up in the marketplace, Chris Blackwell never doubted the ultimate success of the DVD format. He was convinced from the moment he first saw a movie released on Laserdisc.

"There is a film that I put out when I had Island Pictures called *Koyaanisqatsi*. The film was made by Godfrey Reggio. There were no actors in it and no dialogue or storyline—it was just images set to music by Philip Glass. When I saw *Koyaanisqatsi*, a light went off in my head, and we picked it up, released it, and it did quite well. It has become a bit of a classic in a way. From that moment on, I was just waiting for the time when there would be an audio-visual format that was going to gain acceptance with consumers and creators. Soon after Laserdisc came out, everyone started going digital. Laserdiscs hung in there until DVD finally came out, and when it finally arrived, it was something that I had been waiting for, for ages. I think that DVD hasn't even started yet. At the moment, it is still used more as an ancillary for movies. But, DVD is an incredible platform for creative people."

Nurturing Innovation

With a market becoming more accepting of the blending of different forms of media, Palm Pictures keeps looking for new ways to communicate with digital media. "We are always looking," Chris said. "Some things originate from music, and others come from somebody just wanting to make a little film about something. I feel that DVD will create a new kind of business, somewhere between the record business and the film business. Between the two, because I think the costs will be much less. I think that all the stories and elements that you really want to see will be released and marketed on DVD. You will buy them just like you buy a CD. Just like you await the new CD of an artist, you will await the new DVD from this director guy who makes these films and you have no idea how on earth he ever thought them up, or thought about making something in that way."

Future DVD Distribution

The new model for entertainment, Chris believes, will be something that sits in between where the film and music industries are right now. "I don't think it is going to be competitive with the regular film business. There is just so much competition for somebody's leisure currency. The fact that there are cinemas that are going to distribute digitally just means that this one person—this imaginary person that makes something totally unique for $2000—might reach a wide audience. You won't have to spend fortunes blowing it up and making prints and shipping them out; you will just have them distributed on DVD to theaters."

Palm Pictures continues to explore different means of title distribution. One method that appeals to them is the approach used by *Res* magazine. As David explained, "*Res* is an integrated DVD, print, and Web magazine. It uses a different model in that sponsors pay for the disc and the readers get the DVD for free. It is another type of distribution channel that is subscriber-driven. The magazine has two covers: one for subscribers, and one for newsstands. The DVDs are only included for subscribers because newsstands are reluctant to carry anything that is polybagged."

With this more free-format approach to the medium, the *Res* magazine staff can include content outside the realm of typical Hollywood DVDs, including short-form content, travel programming, music by innovative artists, and so on. The success of *1 Giant Leap* has helped to reduce some of the barriers to presenting unique content on the medium. Retail channels, however, still run into problems categorizing content that spans both music and video.

Chris commented, "I think retailers in general are embracing DVD, because it is something fresh and something new that they can introduce through their system, but I still haven't seen any of the retailers figure out how to merchandize it properly. If it is a music-driven DVD, often it is pushed to the DVD film buyer, rather than being integrated with the CDs to capture the music buyer. In a way, if something is a music-based DVD, you really want to have it in the music section, but because we put DVDs in a larger case, they don't fit properly in the racks in the music section."

Nonetheless, Chris sees growing acceptance for DVDs that present music rather than films. "Music-based DVDs are beginning to sell very well," he said, "and they are starting to be really embraced. So, I am sure that you will soon see much better merchandizing, because there will be so many more titles in the marketplace. When there are only two or three, it is not worth it for a retailer to set up a new system for merchandizing. Lots of companies are delving into their catalogs and/or making new DVDs. A lot of new acts, frankly, view themselves as being audio-visual acts. They want their releases out not just as audio CDs, but also as audio-visual DVDs. I think that by the fourth quarter of 2003, you will see much more visibility and much better merchandizing."

Aiming Outside the Mainstream

As Palm Pictures builds a following, the content development typically skirts the mainstream market. "I think when you start anything," Chris said, "you have to start eclectic, at least a little bit. You have to try to build some sort of a base. The mainstream market is very hard to reach, especially for an independent. To get whatever product you are working with in front of the consumer is very expensive."

David also thinks the future is bright for creative innovation, largely because of the availability of a new generation of easy-to-use video and DVD creation tools. "Finally," David said, "DVD has been around enough and has been established enough as a format. And, I think that these content creation tools have come far enough. When somebody gets an idea, they can go soup to nuts without a budget at home. They can shoot on their consumer DV cam. They can edit on Avid DV Express or Final Cut Pro. And, even as dumbed down as iDVD is, they can author their own DVD. All you need is a Mac or a PC with a DVD burner. So, even though you probably don't want to take that route for big commercial releases, this is really giving more people an opportunity to present more interesting ideas."

David continued, "If somebody walked in the door and said, 'I have an idea to make a movie, but I've never done any filmmaking in my life and I don't have any budget. Do you want to invest in my movie?', the average film company's reaction would be, 'No. You have no stars. You have never been to any kind of film school or done anything like this before.' But now people can actually go out and do it. Because they are approaching it without the stringent framework of what the average filmmaker has learned about how to make a movie—not only the process, but securing funding."

"Half the time," David said, "creating content is about finding the money to do it. Now, people have the freedom to really experiment and do things outside of the system. *1 Giant Leap* was done in our system, but I think Chris is probably the only guy in the business who would have enough vision to do something like this. Who else would be willing to say, 'Take these cameras and go around the world and make a record and a film.' And this with guys who had never made a film..."

David thinks that the evolution of digital media is going to get a lot more interesting. "The idea that people are starting to think of visual content in a non-linear fashion is really amazing," he said. "They have digested music in a non-linear fashion for a long time. With the CD, people are so adjusted to that. But, historically, video has been something that you put in the player and you watched it from beginning to end. Now, people are beginning to open up to short form content and other more experimental content. I think the idea of the personalized viewing platform is powerful—the fact that people can watch a movie on their PC, and, soon, on their cell phone."

The big challenge in the industry, David feels, is finding the right content for the right platforms. "When you think that a lot of people have huge TVs at home and go home and sit on the couch and watch the news—you don't need a huge TV to watch the news. I have a cell phone that has video and the video quality is fine for watching the news, but you are probably not going to want to watch a big Hollywood blockbuster on your cell phone. You want to watch that on a real proper system and screen. Also, when someone is at home, they are not necessarily going to jump through short form compilations of entertainment. There are a lot of viewing environments where you only have 15 or 20 minutes—maybe you are on a plane or in a cab and you think, 'Maybe I will watch a short film.' The difficulty is getting the content out there and making people aware of it. The Internet is a powerful tool to help build awareness for projects. DVD is a portable, tactile, sexy format that actually sounds good as well as looks good. VHS never sounded very good. All of these things combined with all these possible viewing platforms are creating a visual revolution. For a lot of content, you might view it in a Starbucks, or you might view it in a cinema, or you might view it on your cell phone. The challenge is matching the right content with the right platform. This is all good for us, because things are opening up, creating much more demand for content. People nowadays are used to stimulus. They want to see audio-visual content everywhere."

Palm Pictures is well positioned to take advantages of these shifts in media use and presentation and their latest projects are pushing the boundaries of the format even farther. Chris says simply, "We're trying. We love DVD."

Asked if he has any words of advice for aspiring DVD developers, his reply is equally concise, "I think that they are in the right place at the right time."

10

Case Study: Making Music on DVD

Over the span of more than thirty years, Stefan Grossman has turned his love for acoustic guitar music into a long-term, home-grown, sustainable business. In the process, through historical videos and guitar lessons, he has taught countless guitarists a wide range of styles, covering the blues legends of the 20's and 30's to seasoned veterans such as John Renbourn and Ton Van Bergeyk. Most recently, Stefan has begun converting his extensive archives of historical footage and video lessons to DVD format, taking advantage of the interactive capabilities of the medium to break up a series of lessons into easily accessible segments using DVD chapter markers.

This chapter explores the evolution of *Stefan Grossman's Guitar Workshop* and the techniques he uses to produce DVDs of both historic performances and note-by-note fingerstyle and flatpicking guitar lessons.

Audio Lessons and the Early Years

The odyssey began in 1964 when Stefan embarked on a journey across America, stopping along the way at folklore centers to give guitar workshops and offer lessons. A network of guitarists developed, many of whom agreed to keep in touch, and from this, a series of audio lessons was produced. This series became the genesis of *Stefan Grossman's Guitar Workshop*.

In 1967, the travel bug compelled Stefan to extend his explorations to Europe, Asia, and Australia and in each location he met many guitarists, some of whom became lifelong friends. Through concerts, lessons, and guitar workshops, he honed his skills, delved into unfamiliar styles, expanded his repertoire, and continued to build his business venture.

"About the time I moved to Europe," Stefan said, "my father got cancer. He has always been a very active man. My mother said to me, 'Is there something you can do to help keep him busy,' because he had retired at that point from his own business. I said, 'Sure.' As a result, I greatly expanded the audio guitar lessons."

During this period, Stefan together with ED Denson had founded a company called *Kicking Mule Records* with offices in both the U.S. and Great Britain. A natural synergy and clear business direction developed through concert tours and friendships with a growing community of internationally known and fast-rising guitarists. As Stefan describes this period, "I was in contact with a lot of acoustic guitar players. Initially, *Kicking Mule Records* just dealt with acoustic guitars. At this time, I was living in Rome, Italy, performing and hanging out with great guitar players. I am a real guitar nut in the sense that I always wanted to learn guitar playing. I thought, I'll combine things. Since we're on tour with Dave Van Ronk, I'll have him and some of the others do audio lessons, if they want to do them. John Fahey. John Renbourn. Ton Van Bergeyk. A lot of guitar players—either I wanted to learn their stuff or I thought it was important historically to document what they were playing."

From the beginning, Stefan had a deep sense of the importance of capturing a slice of music history. "My interest in music," he said, "was always from the 20's and 30's. I always thought, 'Wow, if only we could have recorded more things from then.' It is very important to the moment that you are living in to realize that it is history. And to try to document it—to put it down somewhere."

Capturing Material and Growing a Business

Somewhere in the midst of all the touring, the business started to grow legs and reach an expanding audience of guitarists and those who enjoyed guitar music. Stefan focused on recording the abundant material and decoding the arrangements to create lessons. Back in the states, Stefan's father was thoroughly engaged in setting up a solid foundation for the business and developing sales channels for the audio lessons. "My father got crazy busy," Stefan said. "And he is still alive. It definitely got his mind off the chemotherapy. I was in Europe for 20 years, so he ran the business for that 20 years. And it literally kept him really involved and it was great. It was great for me. It was great for the artists. It was great for my father."

When Stefan returned to America, he decided to take a more hands-on, approach to the business after discussing it with his father, who was initially a bit hesitant to hand over the reins. Technology was to become a major part of the business direction for the next few years.

Technology and Music

Well before the arrival of DVD on the scene, a succession of technologies for capturing sound, recording moving images, and managing business operations changed the ways in which Stefan approached the key focus of the business: teaching guitarists how to play music and relating the history of some of the most important musical forms.

"Technology...," Stefan said, pausing as if to consider the word. "We have always been arms length from it, but then when we grasp it, it really changes our life. The first thing was computers. I never thought that I would get involved with computers. Nowadays, I'm in front of a computer screen maybe 12 to 14 hours a day, doing design work or writing. First, we got a database program and that enabled us to focus on the mail order side of the business and make it more of a business. Then, the catalogs. We could easily put together catalogs ourselves with desktop publishing tools. But, we were still just doing audio lessons."

Teaching through Video

The benefits of video weren't immediately evident to Stefan. A good friend, Happy Traum, had recently announced that he was going to start using video to teach guitar. Stefan was skeptical. "I thought, who would want to do video? Audio lessons are so much more powerful. Especially if you are teaching music from old artists like Reverend Gary Davis, Lonnie Johnson, Robert Johnson, Charlie Patton—all of these old legendary blues men. If you want to teach students how to play like them, it is really fundamental that they hear what those people sounded like. And, you can't do that on video. You can't stop in the middle of a video lesson and say, 'Let me play a Gary Davis piece for you.' Because you would have an empty screen. So, I was reluctant to get involved in video."

Eventually, Stefan decided to try this new approach. "We recorded three videos in a college studio," he said. "I was living in England then and had a friend who was a media professor. We put out the videos and they were a big success."

Hands-On Production

A hands-on production approach suits Stefan. "I'm always involved in all aspects of my business," he said, "whether it is designing the catalog, designing the covers, or handling the production of the videos. The first videos were of me. My friend was the director. I put together three videos and I showed them to Happy. He thought they were pretty nice and said he was interested in putting them out."

Stefan was still hesitant about this new approach, but decided to release the first videos himself. "It was a whole virgin territory," he explained. "One thing led to another and all of a sudden the video lessons became like our audio lessons. We had so many contacts with guitar players who were friends, that we just started recording. With an audio lesson, though, we could do high quality lessons in my living room on a Revox and I could edit them with a razor blade (this is way before the era of digital editing). Guitar instrumentals can be complicated, so you have to do a lot of editing. With video, it is a different world. I was not acquainted with that world, so all of a sudden, whereas I had been doing the engineering and the editing with the audio lessons, I had to depend on other people for doing the lighting, doing the online editing, and so on. That was a little bit of an adjustment."

Rising Production Costs

Each new advance in the technological realm seems to have a whole new set of expenses, as Stefan quickly discovered. The video lessons were significantly more expensive to produce than the audio lessons and each step up in quality brought a whole new set of costs. "We started cheaply," Stefan said, "but as time went on, we kept improving our production quality—starting with better cameras, going from Umatic up to Beta SP. Now, everything we do is on Digital Beta with much better cameras. All of a sudden, there were much bigger bucks involved in production."

Despite the risk involved in adopting the higher-end production processes, the new video titles proved successfully. "Yeah, it was bigger risk," Stefan said, "but it always paid off. Everything we did in video was always able to break even and some titles were really successful. We were able to do a lot of things that I thought were important "historically." I knew we would sell the 300 copies needed to break even as long as we were tight with our costs."

Early DVD Experiments

Although *Kicking Mule Records* was still operating, a partnership dispute led to the dissolution of that business. An equitable agreement was reached whereby Stefan kept the masters that he had been involved with and his partner kept those that he was most closely associated with. With this library of archive material, Stefan returned to America in 1987.

"Shanachie Records brought me over to America," Stefan explained, "and I worked there for five years. While there, I put out a series called the *Guitar Artistry*—that was at the time when vinyl was disappearing and CD was coming about. At Shanachie, we decided that instead of just reissuing the *Kicking Mule* stuff that I owned as it was, we would reissue a series of CDs, a *best of* series. We would, for example, take three Stefan Grossman LPs and make one CD that would combine the best tracks of the LPs. This series, *Guitar Artistry*, is still available."

Back to Video

"Meanwhile, on the video side," Stefan continued, "we kept doing videos and trying to improve on the lessons this way. At Shanachie, one of my jobs was to start their video catalog. We knew that DVD technology was coming about and we thought this was going to happen in 1988. One of our ideas was that if we put out a video of Muddy Waters, everyone would buy it—right? They would have to buy it. Or, videos of old blues men. Or, Wes Montgomery. If Wes Montgomery sold 100,000 albums or CDs, he would undoubtedly sell 80,000 videos. Right? Wrong—that's not what happened."

As everyone who followed the early stages of DVD technology, it took a very long period—several years—before the right balance of inexpensive players and abundant choice of titles fueled the explosive growth rate. Once that growth became evident, however, sales channels opened up that were previously unavailable. DVDs were less cumbersome and took up much less shelf space than video cassettes and an audience enamored of the audio and video quality of this new medium actively sought out new titles to buy and play in their new acquired systems.

Stefan had tried for years to gain wider distribution of his video titles through retail outlets and shops with limited success. Suddenly, with DVDs achieving tremendous popularity, distributing titles became much easier.

"With the video titles," Stefan said, "the shops just didn't want to know. The wholesalers didn't want to rack videos. After five years, I left Shanachie on good terms. I decided to start a series called Vestapol, which would be like reissued old records, but instead it would be reissued footage. I could produce titles like the *Legends of Country Blues Guitar*, an anthology of all the footage I could find. It was exciting trying to locate the footage and license it and find material that people didn't even know still existed. We started a whole series of historical videos, and, because they were so successful, it was a self-financing venture. Stores would take the historical titles more readily than instructional stuff, but still with hesitation. With the extra money, we could go and produce concerts (Tal Falow, Herb Ellis, Charlie Byrd in concert, Bob Brozman in concert, Pierre Bensusan)."

Changing Financial Stakes

The production considerations were not inconsequential. Stefan was weighing financial stakes that had escalated from $50 for producing audio lessons to $2000 for producing an instructional video in an afternoon to the new reality of DVD. "It was now getting to be five to ten thousand dollars to get a whole crew out there to record the material, including three or four cameras with a gib."

Throughout the recording and film industries, many companies seemed to be waiting for some signal that DVDs would succeed in the market, and during this time a level of caution often influenced business decisions. Stefan began talking to some of the industry players to see the level of interest in his historical titles. "At the time," Stefan said, "I did approach Columbia, Warners, and others, saying, 'Hey, I've got this library—stuff with Wes Montgomery, Chet Atkins, Merle Travis, Barney Kessel, Joe Pass, superstars.' I felt I needed funding to get into DVD—it was very expensive in the beginning. But, they all passed. No one was interested, which is, in fact, fascinating. If you look around today, those companies do not have any catalog of DVDs. They may have new products or new products, which would be the pop or the rap music. But, if you look at Columbia, they don't have the material like they have in their Legacy series of CDs. I was always fascinated why they didn't bite. I knew a lot of the execs there. Some of them were very excited about DVD, but as it went higher up the food chain, the people at the top just didn't see it as a viable thing."

Stefan sees it as a responsibility to keep our musical heritage alive, even if many of the executives in the recording industry don't share the same vision. "It is responsible to put out the historic recordings of Duke Ellington," Stefan said. "It is not that expensive, so you don't need to sell 500,000 copies. You might only sell 5,000 copies, but I still see it as a responsibility."

The Bootstrapping Approach

The entry price to the promising world of DVD was still high and Stefan turned to a friend, Richard Nevins at Shanachie, for advice. He asked Richard how to deal with the high expenses involved in DVD production. Releasing several titles into this new market would require more cash outlay than *Stefan Grossman's Guitar Workshop* had ever spent at one time. Richard's reply was instructive. "He said," Stefan recalls, "No, no, Stefan. In our experience, you put out two and you get them out into the wholesale market. Then, the money you earn from those two titles in the first 90 days will enable you to do another two. And, then you do another two. And, another two." Stefan pauses for a beat. "And that is exactly what I have done," he said.

This bootstrapping approach has served his business well. "We are totally undercapitalized," Stefan explains, "which is both good and bad. It means that when we do things, we have to be very tight on our inventory and we have to be very tight about what we do. The business has to pay for itself. We are never in a position of saying: 'Oh, here is $500,000—let's put out 50 DVDs now.' This, in fact, would be bad, because the wholesale market wouldn't be able to take all of those titles to the distributors. So, we started to put out DVDs and they are distributed by Rounder. I have an arrangement with Rounder and with Mel Bay Publications. Mel Bay Publications does very well with our instructional stuff, and also with our historical stuff (which has always done well with them). Rounder does very well with the historical stuff, but they do not handle the instructional titles. Once we put out the DVDs, all of a sudden, the shops were buying them. The same shops that never bought the videos were now racking the DVDs."

This newfound acceptance of the medium has definitely expanded available distribution channels. Stefan thinks the change is due to both a combination of the lesser shelf space required by DVDs, as well as the rapidly growing popularity of the medium. "When Disney went to DVD," Stefan said, "all of a sudden the shops started to say, 'Well, it's happening. And videos aren't. DVD is happening, so let's get into it.' But, what was also fascinating was the rate of sell-through. They would put DVDs in the racks and customers would quickly buy the stuff. Suddenly, we're in a situation where Rounder used to order inventory of the VHS videos in quantities of five or ten, now they're buying 300 or 600 copies of the DVD."

The Future of Guitar Instruction

With the discovery that the DVD market might prove an entirely different animal than the video market, Stefan has re-examined his entire business model and the titles in his catalog with an eye to the future. The early success of the historical DVDs makes him wonder just how to approach the archives of instructional material. "The Vestapol titles are really doing very well. What should we do with our instructional stuff? We had about a hundred titles in Vestapol and we've come out now with about 50 titles in DVD and we're coming out soon with another ten. We're getting to the point where some are not such big sellers and some that are licensed can't be released as DVD." As the historical titles have largely been through the conversion process, Stefan wonders how best to handle the large library of instructional content.

The Right Format

DVD has proven popular and successful in areas that VHS video failed, as Stefan has been discovering through a series of experiments. "We had a John Renbourn retrospective video from 1965 to 1995, which I thought was great. John Renbourn is very well-known. When the video came out, it only sold about 300 copies. I thought, 'Let me put out the DVD and we'll see if it is a dog, or not.' We put out the DVD and it sold about 3000 copies within a month. It just shows you: it is a question of getting material on the right format—DVD—and you're off and running."

The question of how many copies of a title can be run and still become profitable becomes more pressing as the instructional titles are considered. "With DVD," Stefan said, "when you do a run, you have to do a minimum of 1000 items. I know I can sell a thousand of anything of the Vestapol historical titles. But when I consider the instructional material and I have the music of Turlough O'Carolan as taught by Duck Baker, and it has only sold 200 instructional copies, do I dare put it on DVD?"

Both VHS and DVD

Straddling both the VHS and DVD formats adds inventory difficulties and complexities to the mix. Stefan actively questions whether newly released titles should be produced in both formats. At the moment, he is leaning towards an all DVD future. "From my discussions with Happy Traum and my other peers in the business," Stefan said, "DVD players have come down so much in price that if a person is really interested in buying a $29 instructional video, they can go out and buy a $100 DVD set-top player. In fact, maybe it will force them to do that. For us to put out that same title in VHS format means a lot of money tossed in the wind. It may break even, but just with inventory, who needs those problems? Is it worth the gamble just to do DVD? The reactions I'm getting from email, almost 98 percent of the people are saying to go with DVD only. People who don't now own a DVD player are saying that they would buy one to view the lessons."

Duplication or Replication?

Stefan has considered using a DVD/CD duplicator to handle short runs of instructional titles that may not generate sufficient volume to justify the minimum run of 1000 replicated titles. "We have audio lessons that we are converting to CDs," Stefan said. "The ones that we thought might be too esoteric, we thought that maybe we would do them inhouse. And, we still might do that. Again, the technology is just exploding. Initially, you had to author DVDs in a proper studio. Now, with Apple's system, you could conceivably author at home. We don't do that yet, but we have the capability to do that. I'm discovering, though, that people who have older DVD players sometimes can't play new DVDs."

It's worth noting that newer DVD players that are MultiRead-compliant can typically handle 95 percent of recordable DVD-R and DVD+R discs and CDs, but this still doesn't fully resolve the problem of compatibility with earlier generations of players or that 5 percent of the audience that has compatibility problems.

Setting up an inhouse production facility for DVDs is affordable and makes sense in many circumstances, but sometimes relying on professional engineers and high-end equipment can be more practical for a project. "When we were recording records," Stefan said, "the big question was: shouldn't we build a studio? Why should we pay a studio to do our stuff? If we build a studio, it will save us money. In 1967, I thought this was a great idea, but then what happens when stuff breaks. You need to get the engineers in to be able to fix things. Or, you want to keep upgrading. So, what is happening now, whereas it cost about 10 to 20 thousand dollars to make a DVD two or three years ago, now it costs about 10 percent of that."

Streamlining DVD Production

Though many of the DVD producers discussed in this book's case studies prefer to handle the end-to-end process, from videocam capture through DVD authoring, good reasons exist in some circumstances to give portions of a project to the specialists. Stefan has found it more efficient to use a studio and video engineers to create the instructional lessons on DVD. The reasons that apply may also make sense for other DVD producers who have unique circumstances. Sometimes, inhouse non-linear editing adds unnecessary steps to a process that works quite well in a conventional studio setting and professional editing facility.

"When we do video editing for our instructional stuff," Stefan said, "everyone says, 'Well, you can use Final Cut Pro at home and put it together on your own system.' In my case, this doesn't save me time. It only takes me four hours to do instructional video editing, because of the way we do them. We are using a top studio in New York. We get the dead time hours—they give us pricing that is very low."

Being flexible and scheduling projects to fit into openings in the studio's time slots can make a significant difference in the overall cost. "We do most of our work inside normal business hours," Stefan said, "because the studio knows that we are a steady client, but the business is changing. We used to work at a place that dealt with advertising people. They were paying $450 an hour. We were paying $110 an hour."

This approach still gives Stefan the benefit of top talent in the field. "We are getting the top equipment and the top engineers. We would typically go there on weekends or when no one else was working (the studio would ask us if we wanted to come in and do some work). The boss would sometimes allow the engineers to do their own work on the weekends and use the equipment."

Stefan has since located a new spot for video work "Now, we are at a place which is an old proper studio," he said, "a major studio, but we get very good rates because they know we can fill in slack time. We do work there from 10 in the morning to 6 at night. We don't do it seven days a week. We do it perhaps twice a month."

Advantages of a Live Switch

One of the advantages of working in a professional studio setting is the availability of multiple cameras and an editing console where it is easy to perform a live switch. "When we do an instructional title," Stefan said, "one of the important aspects of the process is to be able to do a live switch. You have three cameras available. We're not doing iso's. It is sort of like doing a live television show. Doing it that way lets you get much more energy on the screen. The person teaching really has to have his shit together. And, the cameramen, who are all guitar players and all work at CNN, are all top-rate people—it becomes like an organic unit. In a ninety-minute segment, there may be one switch that is wrong, that could have been better. But, it is not worth the difference than if we had iso's and had to go online—it would take hours to edit that show."

"With a live switch," Stefan continued, "when I go online, it just an assembly edit, just putting together all the pieces. The guy plays the song. It dissolves to black. Comes back up. He talks about the song. A ninety-minute session only takes three to four hours to put together. It is illogical to do that on a non-linear system. It is better to do it on the old-fashioned linear system. We don't use Avids or Final Cut Pro for that process. It would be stupid to put all that material into the machine, to edit it, to put it back out of the machine."

The conversion from format to format would add a number of cumbersome steps to a process that Stefan has refined quite successfully using professional video editing tools. The nature of the content and the ability to carefully choreograph the videotaping makes much of the difference. Stefan considers what it would take to adapt this same process for non-linear editing. "With Digital Beta tape, I would have to pay studio time to have them transfer the video content to a hard disk. Then, I would go home with my portable hard disk, do the editing in Final Cut Pro, and get it out. Go back to the studio, and they would then have to put it on Digital Beta for the duplicator and for the person who is going to author the DVD. The studio time just for that is going to be three hours."

The DVD authoring in this process relies on providing a digital master of the video and a time-coded VHS cassette to the person handling the authoring. Stefan emphasizes that the nature of his content doesn't require complex authoring work. "I always think of our stuff as 101 DVD Authoring or 101 Editing. It is not so complicated. With the historical titles, it is just like a CD. The authoring is pretty simple: someone needs to know when the track begins and the track ends. Likewise, with the instructional videos. In the historical stuff, there aren't multi-camera views; you are fortunate just to have what you've got. If you have Wes Montgomery from 1965 in a kinescope, that's all you've got."

Stefan keeps watching the technologic progress of digital video cameras and desktop video editing tools and DVD authoring packages, but it still doesn't make sense for him to alter his process. Having access to some of the best engineers in the business can make a big difference in the overall quality of the video content. "One of the engineers that I work with is one of the top engineers in New York. What takes him three hours to do would probably take someone else eight hours to do. It is so much nicer having that responsibility on his shoulders rather than mine. When I've priced it out, there is so little savings that I would rather do it the way I'm doing it. Delegate the video work to other people."

Reviewer's Expectations

Stefan used to send out newly released DVD titles to reviewers, but he quickly learned they tend to focus on the bells and whistles rather than the content. "Right off the bat, we discovered that they're not into the music, they're into the format, asking questions like, 'why doesn't this DVD have everything that *The Matrix* has?' Our clientele is not even interested in that. That want to know how to learn this guitar piece or they want to hear Wes Montgomery play. Period. You do see a lot of Hollywood films that are similar, just putting the film onto DVD and adding chapters. If they have an interview with the director, they include that. But, even they get slammed in the DVD reviews. Someone will say, 'hey, this doesn't have all these other bells and whistles that some of these fancy guys have."

For future DVD projects, Stefan may incorporate some of the more interesting aspects of the medium, such as the ability to accommodate multiple camera angles. "Now," he said, "if we were doing a concert filming, where we used use a live switch, we would do a three-camera or four-camera iso. We would spend the extra money, so that when we put it into DVD, we would have all those different camera angles. Whereas, if you're going into the studio online and you already have a live switch, there is nothing you can do with camera angles on the DVD."

Putting Concerts on Disc

Stefan Grossman often gets involved in the acoustic guitar concert scene, sometimes partnering with other parties and sometimes independently producing concert video for conversion to VHS and DVD. The projects are sometimes planned and sometimes arise spontaneously in conversations with friends and acquaintances in the business. As Stefan explains it, "When we've done things like Bob Brozman, John Renbourn, and Pierre

Bensusan in concert—those were artists that we know and they were going to do concerts in clubs that we knew. They would come to us in some cases and, likewise, in other cases, we would go to them and say, 'Let's record the concert.' We would then arrange it with the club and get a crew to come in and do the shoot."

The first few songs from one recent concert, starring Bob Brozman and Ledward Kaapana, appear on the bundled DVD included with this book. Stefan talked about how that project originated. "It was an interesting scenario how that came about," Stefan said. "George Winston and I keep in touch. He is a blues player, besides being what he is famous for. He does John Fahey stuff and country blues. So, I have always been sending him our stuff. He said, 'Bob and Ledward will be playing in Pittsburgh' and I said, 'Let's try to organize it.' We paid for the actual taping; it was almost like a joint venture between us and Dancing Cat [the recording company that releases many Brozman/Kaapana CDs], but we released it. So far, Dancing Cat hasn't gotten in to the visual world."

The Manchester Craftsmen's Guild proved an ideal venue for the concert and the logistics of taping live guitar in the well-equipped facility had been refined to such a point that Stefan didn't feel that it was necessary that he oversee the taping process. "When we did Larry Coryell instructional videos at the Craftsmen's Guild, I was there. But when we are doing concerts, I can just leave it in the hands of Marty. I realized that Jay and Marty have their act together so solid, that if I was in the control room, all I would be is a silent witness. I wouldn't even want to get in the way of what they were doing—they really have it all down."

From VHS to DVD

When Stefan returned to America from Europe and began working at Shanachie, he had a clear sense that DVD would be somewhere in the future and even began visualizing how projects might appear in the new medium. The thinking at the time was that this digital media would be similar to CD but with added video elements. But, early on there was doubt about the technological advantages and the introduction of DVD kept getting pushed further and further into the future.

As he began achieving success with his own VHS titles through *Stefan Grossman's Guitar Workshop*, Stefan still didn't think DVD had sufficient momentum to risk the investment. "The first two DVD titles we actually licensed to other companies to do," he admitted. "We didn't know what was going on. There was a big fight about the format. I thought, 'What is going on here?' Here is a great format and instead of just saying, 'here is the format,' these guys are saying, 'here is the format and now here are five variations of it,' just confusing the situation."

At this point, however, the popularity of DVD is clearly soaring, even though VHS users are still pretty active. As Stefan describes it, "In this last year, I have really gotten the feeling there is impetus behind DVD, but we were just having a summer sale here and I'm just amazed at the number of videos we are selling. My attitude is: our new releases will only be in DVD format. I am presuming that is the right business decision to make. But, boy, we still get tons and tons of video orders. Not from wholesalers, but from the actual customers. You can almost tell with our clientele their format preferences. We have the Chet Atkins stuff for the country people. And, Doc Watson. The jazz people—the jazz people are more into DVDs than the Chet Atkins people."

More on the Production Process

Working with Rainbow Video in New York City, Stefan has devised an end-to-end method for handling the production process, moving from DigiBeta tapes to DVD test copies to replicated DVDs. One reason the process is so successful is the long-standing relationship with Ron Perlstein at Rainbow, who oversees the many details involved in creating the DVDs (for more on this process, see *Working with Rainbow Video* on page 131).

Stefan describes the production process in these words. "After videotaping a concert or other project, I've got a live switch of everything. I then get back a timecoded VHS from Rainbow so that I can check out what is going on. If I'm at a concert while it's being taped, I would have been writing down the timecodes as it was happening. But, because I wasn't, I used the time-coded VHS tapes. Then I establish a map on where I want a segment to be sitting on a video DVD."

"For example," he continued, "the timecoded VHS will include both the interviews and the concert footage. There will be different reels included on it. You just say, 'I want the *Summertime* performance and here is Herb Ellis talking about his arrangement of *Summertime*.' You have a list of all of these times and when you go online, there is no sense to do that. Because you already know all your ins and outs, it is a kind of meat and potatoes approach to editing."

The timecodes identified during the creation of the final master tape provide the in and out points for the DVD authoring, as well. "We get our master tape at the end of the day," Stefan said. "After the online editing has been completed, I get a VHS timecode of that. So, I know where the chapters should be—where they should enter and what they should be. That is the information that I give to the DVD authoring person."

Instructional and Historical Booklets

Almost all of our instructional or historical titles that *Stefan Grossman's Guitar Workshop* offers come with booklets. Along with the timecode of the title, the authoring person also receives a CD-R that contains the Acrobat PDF file for the booklet. When someone inserts the DVD title into a computer, the program content is available as well as a folder that contains the PDF of the booklet. The booklet can then be read onscreen or printed out.

"The booklet," Stefan said, "also will have biographical material. It will have photographs. It will sometimes have music. Usually this content would have been in a printed booklet. I would love to get away from the printed booklet completely, but people like to have it when they open up the DVD. They say, 'Oh, we have a nice 80-page booklet. How cool.' The problem is: what happens when they lose that booklet? Originally it was downloaded from our site, because we didn't realize we could put it onto the DVD. Now, it is on the DVD and you get it through your computer."

The minimum print runs for the booklets were another factor that led Stefan to rely more heavily on using PDF files for the materials that are otherwise printed. "The printed booklets will be included with the first 2500 DVDs," Stefan said, "and then after that, there is no booklet. We have not gotten one complaint, whether people get a printed booklet or a PDF booklet. It doesn't seem to matter. Our tendency is to give the customer something extra. But, it doesn't seem to really matter, quite honestly, to the customer. Customers seem to be very happy with either format."

A side benefit of providing the booklets in PDF format, they can be printed out on 8.5x11-inch paper, making the music much easier to read for guitarists. In comparison, the printed booklet measures only 4x7 inches to fit into the back of the video box or into a DVD box.

Musical Notation

"We use Finale for the musical notation," Stefan said. "But, as far as the tablature is concerned, almost every musician teacher that we work with has their own idiosyncratic tablature. So, we do not standardize on the tablature. We want that to reflect how the teacher looks at the music—it is a very personal thing. The instructors supply the tablature and we don't dictate to them what system they have to use. Whatever system they feel best conveys the music, that is what we go with."

Test DVD

As the DVD production process continues, Stefan typically receives a test DVD from the person doing the authoring. He then checks to ensure that all the ins and outs are in the right position. He also has a strong aversion to viewers being forced to watch the FBI warning at the beginning of a DVD. "I made sure" Stefan said, "that as soon as our DVDs start up, you can press menu and go right to the menu. You don't have to go through an FBI warning or through trailers or anything else."

Once the position of the ins and outs has been confirmed and Stefan double checks to see that the PDF booklet file is on the DVD, the replicator receives the go ahead to start manufacturing. A certain number of titles are typically kept by the replicator—currently Rainbow Video—to handle order fulfillment, an extra service that Rainbow provides to simplify Stefan's logistical needs for order processing.

DVD Plans for the Future

Initially, Stefan plans to convert his entire catalog to DVD. One problem is that many of the earlier titles are stored in Umatic format and Stefan is concerned that the video will not hold up well when converted to DVD format. "The Umatic tapes," he said, "are softer. They are older tapes. You have dropouts. You have a lot of technical problems. They weren't created using digital cameras. The equipment wasn't as fine."

Many of the artists shown on these earlier titles, such as Dave Van Ronk and John Fahey, have died, so this earlier material becomes part of the historical record of their musical accomplishments. Stefan plans to work closely with Rainbow Video to transfer much of this earlier video, using automated tools that can enhance the contrast of the images. "The software use," Stefan said, "is sort of like Photoshop. It can enhance the contrast by making the pixels harder. That improves the contrast, but will the program create a problem for each problem it solves? We don't know yet. With some of the earlier stuff, we might put a disclaimer on it, like they have on Yazoo Records. If you get a recording of a CD of Skip James' 1928 recordings, it is going to be scratchy. Take it or leave it. We can't do it over. These early artists aren't here anymore."

Irreplaceable audio, video, and film recordings can be restored in many cases, and it is an open question whether artifacts and flaws in the original materials add to the sense of history and simply degrade the presentation. "A certain part of the customer base," Stefan said, will say that the flaws increase the historical sense of the material. But, I know that when I watch a Ken Burns documentary, which has millions of dollars going into production, and there is a crease in a photograph, you wonder. Why didn't they go into Photoshop and take out that crease? It would take them maybe half an hour. That type of thing. When someone gets a DVD and all of a sudden there is a glitch or a dropout, they may think historically it is important. But, hey, isn't this a technical problem? You have to have a disclaimer saying, you will get dropouts."

"One of our Vestapol historical recordings of John Lee Hooker," Stefan continued, "contains some footage that is rough. We let customers know that is rough and we've never had a complaint about it. Whereas, with instructional material, people may think a dead artist is still alive. When we start converting our instructional titles, one of our biggest sellers— the first video that we did—was called *Fingerpicking Guitar Techniques.* Since I did the original guitar work, I was able to rerecord it. The new version will be from DigiBeta source material and it will be as sharp as anything in our catalog."

Stefan enjoyed returning to the studio to recreate this milestone work. "It has been such a huge seller. Here I am 20 years older going into the studio to see if the songs come out with the same feeling they did originally. Obviously, it is slightly different. I was very pleased by the way it came out."

"I'm starting from the best sellers," Stefan said, "and working through the less recognized ones. I also have to have the cooperation of Ron at Rainbow. We have two catalogs. One sells pretty well and it finances itself. Now, we are going to this other catalog. We know that it is maybe only 10 to 20 percent as commercial as the historical stuff. A title with me teaching guitar is not as commercial as Wes Montgomery playing guitar. When Wes Montgomery is playing guitar, any jazz buff wants to see that. They don't even have to be a guitar player. Someone who buys *Fingerpicking Guitar Techniques* wants to learn fingerpicking guitar—it is a much smaller audience. The problem that we face is whether we can get the DVD authoring prices and replication prices to reflect the difference. Rainbow has been very cooperative, working together with us to help do that."

Documentaries

Stefan doesn't actively produce music documentaries in the style of Ken Burns, but some of the titles in his catalog have a documentary flavor. "In an instructional sense, we have a video where Lightnin' Hopkins is being taught. We have a lot of footage of Lightnin' Hopkins playing. In this title, Lightnin' Hopkins plays a song, the teacher explains it, and then teaches it. The approach is really powerful. The teacher doesn't have to pretend to be Lightnin' Hopkins and sing a Lightnin' Hopkins song (which is pretty difficult)."

"As for documentaries," Stefan said, "we have a Brownie McGhee project on the shelf. I interviewed him about 10 years ago and I have ten hours of interviews and films of his scrapbooks, which were incredible. It is all just waiting to be put together as a documentary. So, this is one project that is on our agenda that we will probably inhouse with Final Cut Pro and non-linear editing. The process is just so time consuming. But, it is something that we think is very important historically. We will do it at some point."

Another Vestapol title has a documentary feel, a series coming out on DVD that was earlier released on video called *Fingerstyle Guitar: New Dimensions and Explorations*. This title includes three hours of material presented as an anthology of different guitar players. "I have always felt," Stefan said, "that our videos and DVDs should be like a visual CD—that our customers did not want a Ken Burns things. In other words, they wanted 60 minutes of solid guitar playing. They don't necessarily want to hear the guy talking—they want to hear the guy playing. So, on *New Dimensions* we first put out 60 minutes of solid music and a booklet."

Introducing New DVD Titles

Live concerts can provide a valuable source of DVD titles and Stefan looks forward to being involved in more of them in the future, but he is determined at the moment to complete the conversion of his catalog to DVD format, a process which is consuming a fair amount of funds. He is also aware that the introduction of titles in DVD format should be handled in a thoughtful way, rather than just flooding the market with new material as quickly as possible. "One thing that I learned when I worked at Shanachie—it was hammered into my head and it was spot on—Richard, the head of the company, warned about competing with yourself. Shanachie used to put out the Chieftains' records. Each one was numbered, one, two, three. After a time, whenever Shanachie put out a new Chieftains' record, they were actually competing with themselves. The shops are saying, 'should I buy the catalog stuff of the Chieftains or do I buy the new Chieftains' record?' I have a feeling that as we keep putting out more of our titles, in a way the titles start to compete against each other. When we put out the first ten, there weren't any other jazz collections or blues collections out there on DVD. Now we come out with more jazz and someone says, 'should I buy that one or buy this one?'

"You would hope they would buy both titles," Stefan continued, "but that is not what happens. You are giving the consumer more choice, so your sales per item may go down as you get your full catalog out. Fortunately, our strategic approach is sound. If you look at our catalog, we are guitar—period. Very little electric guitar—it is mostly acoustic. In Vestapol, we have the jazz thing, with Joe Pass, Wes Montgomery, Barney Kessel. There are eight titles. Then, you have the Doc Watson set. There are seven or eight titles there. Then, you have Chet Atkins and Merle Travis, six or seven. You have the blues, six or seven titles there. You have Freddie King, four or five titles there. Within our catalog, we have a lot of different areas. We do get customers who only buy the jazz or only buy the Doc Watson or only buy the Chet Atkins. But, they don't compete with each other as a regular catalog might do. In other words, if we were just putting out blues titles or jazz titles, it would be a little bit rougher."

Working with Rainbow Video

In many ways, having an established relationship with a DVD replication service can eliminate many of the potential problems that crop up from title to title. A mutual understanding of the needs and working methods of both the client and the service provider can also streamline the production process and minimize obstacles and confusion.

Stefan has such a long-standing relationship with Ron Perlstein at Rainbow Video in New York. "We do a lot of video work with Ron," Stefan said. "He was our video duplicator and he obviously has now gone into DVDs. We have a great relationship with him. He does the replication for us, as well. He does the warehousing. He does our drop shipping to wholesalers. Rainbow may not have the cheapest prices in the business, but it is the best price for the service that he supplies to us."

The DVD catalog conversion goes on even while Stefan is traveling. "Ron is working on 11 titles at once for me. I'm off to England for about 6 weeks. I have to have a whole bunch of stuff ready for September/October release. We have 8 new titles, new recordings of digital instructional material that I'm working on; 4 of them are mine and 4 are Fred Sokolow, who is also a teacher. We record the instruction pieces and then I edit them. Then, we have to take the VHS copies and write out all the music."

The production process starts with Stefan turning over DigiBeta source material and a timecoded chart based on a VHS copy. The chapter divisions for the DVD can be expressed through timecodes once the master has been previewed. Rainbow runs the DigiBeta source tape the whole way through, performing a quality control check. While viewing it, they create the visual timecode, making notes along the way, and then send the visual timecode VHS to Stefan. Stefan specifies the chapter points at appropriate locations. When Stefan gives Rainbow the OK on the master, Rainbow loads the digital video into a computer and begins encoding it to MPEG-2 format for inclusion on the DVD.

One or more check discs are generated of each authored version of the DVD title and each of these is submitted to Stefan for approval. After that, Rainbow makes a digital linear tape (DLT). The tape has exactly the same contents as the test disc except the files get transferred to the DLT.

Order fulfillment for a smaller producer, such as *Stefan Grossman's Guitar Workshop*, can be more challenging than dealing with large-scale producers, but this is one aspect of the service where Rainbow Video outshines the competition, provide a highly individualized service.

More details about Rainbow Video and their services can be obtained by calling 212.594.0545. More information about DVDs available through *Stefan Grossman's Guitar Workshop* can be found on the Web at: *www.guitarvideos.com*.

11

Case Study: Optimal Video Compression

Ben Waggoner lives and breathes video compression, so much so that he has no hesitation at billing himself as the *world's greatest compressionist*, a title that he presents with a wry, tongue-in-cheek twist. Ben has built a career around in-depth knowledge of the art and science of video compression, familiarizing himself with hardware and software encoders, the artifacts that result from poor encoding, the underlying technical details of the codecs, and the techniques to achieve optimal results. His book on this topic, *Compression for Great Digital Video* published by CMP Books, has sold consistently to an appreciative audience and has recently gone into a second printing. From his home base in Portland, Oregon, Ben writes for a number of industry publications, including *DV* magazine and speaks regularly at seminars around the country, sharing his knowledge with anyone out there seeking the most effective ways of storing high-quality video on DVD discs.

This expertise has been acquired through a circuitous route through the industry, leading to the founding of *Journeyman Digital* in 1994 where he began actually making a living from the fledgling video compression industry. These skills were later transferred to Terran Interactive, where Ben founded the consulting services division, and then to Media 100, where he became involved with workflow automation services. At present, he operates his own consulting and services business, providing training and high-profile video encoding work to clients around the world. His seminars and presentations are popular at events such as NAB, DVExpo, and QuickTime Live.

Compressing a Difficult Stan Brakhage Sequence

With a reputation for exactitude and an eye trained for video aesthetics, Ben is often consulted about challenging video compression projects and then hired to get the best measure of quality out of a difficult video sequence. Such was the case when he was hired by the Criterion Collection (*www.criterioncollection.com*), a company that specializes in reissuing classic film features on video, with a DVD collection that spans Alfred Hitchcock to Akira Kurosawa.

Criterion caters to an audience that insists on videophile quality in each release and devotes considerable effort when converting existing video content to MPEG-2 format for DVD release. For a recent project, Ben was asked to compress video segments for a retrospective on the filmmaking career of Stan Brakhage. Without concern for the length of time the required encoding would take, Ben turned to software encoding and produced video segments tweaked and refined to meet Criterion's expectations and to produce the level of quality that DVD viewers interested in film classics expect.

Interested in storing more than the typical two hours or so on a single DVD disc? If you have an application that requires several hours of video and you don't want to go to multiple discs, you can encode the video content in MPEG-1 format. The DVD-Video standard supports use of MPEG-1 files and this technique lets you store up to 40 hours of video on a single disc. The quality, naturally, is not comparable with MPEG-2 content, but you also have the option of combining both MPEG-1 and MPEG-2 on the same disc. "I've seen a couple of cases," Ben said, "where the DVD-Video disc has the main feature in high-quality mode and then includes MPEG-1 files on other parts of it." Most set-top players will support DVDs that include both MPEG-1 and MPEG-2, particularly if they are of reasonably recent vintage.

DV As a Source for DVD

The popularity of miniDV camcorders has spurred a consumer and prosumer revolution. How well does the current generation of miniDV equipment work for providing video files for DVD? The marriage is not exactly made in heaven, as least if optimal video quality is your goal. As Ben explains, "I'm not a big fan of using DV as a source for DVD. The issue is not so much the compression as the color space. DV in NTSC uses what is called 4:1:1 color space. Each color sample is 4 pixels long and 1 tall. MPEG-2 on DVD uses 4:2:0, where each color sample is two pixels tall and two pixels wide. When you convert from 4:1:1 to 4:2:0, it is a mismatch and you wind up with in effect 4:1:0, or color blocks that are four pixels long by two pixels tall. You lose a lot of color detail using that combination. It is not an issue in PAL. DV in PAL is 4:2:0, so it is a better combination."

This color detail loss represents a concern when trying to maximize the quality of video intended for DVD. "Inevitably," Ben continued, "when you're doing the DVD and you use 4:1:1, you only have 180 color samples horizontally across the entire video, but 240 tall. With MPEG-2 files on DVD, you can support 360 wide, but only 240 tall. This means that anywhere in the video image that you have a lot of detailed colors or sharp edges between colors, you're going to lose softness. I wouldn't want to do a lot of 60's style, really bright video with strong colors. DV is a good authoring format, but not the best DVD production format. It is more of a high-end consumer and industrial format. But, you're not going to want to do any kind of intensive motion graphics or any kind of prime-time production work on DV."

While established filmmakers continue to experiment with the DV format, there are limitations to what can be achieved given the characteristics of the medium. Steven Soderberg made extensive use of Canon XL1 miniDV camcorders in his feature film release *Full Frontal*. But, the color loss for motion graphics in particular makes the DV format less than ideal for high-caliber professional work. "Even though he shot it using DV," Ben said, "You can bet that he didn't do the motion graphics on miniDV tape. You can find movies out there shot to miniDV cassettes, like *Blair Witch*, but typically there are higher end processes being used in post-production to finish the work. Filmmakers almost use miniDV as a way to stylistically introduce a kind of degradation into the project."

The higher-end professional digital video standards offer benefits for serious production work. "I think," Ben said, "that DVCPRO-50 is a good choice for high-end work. Or, Digital BetaCam is away great to work with. These formats use 4:2:2 sampling instead of 4:1:1."

Choosing a Rate for Video Compression

Throughput of video data and decompression efficiency are key factors during playback of a DVD. As a part of the encoding process, with the majority of encoders, developers have the option to select the appropriate bit rate. To ensure flawless playback on the widest range of players, these bit rates can be set to higher values for DVD discs destined for manufacturing (replication) as opposed to those being recorded to DVD-R or DVD+R media (duplication).

Ben sees it this way, "The specifications give you a pretty good rule of thumb. You can peak out to about 9.6 megabits safely on most players for a stamped, manufactured disc. Typically, the rate is an average of 5 megabits to a peak of 9.6. This works out pretty well when you are using a good encoder with a two-pass variable bit rate (VBR); the results are generally pleasing."

"One place that people get messed up, though," he continued, "is when they a making a DVD-R disc. Especially with the older players, they can't necessarily read off a DVD-R as fast as they can from manufactured media. So, you have to lower your peak data rate substantially for better compatibility."

The trade-off between compatibility and video quality (as determined by the amount of data supported by the choice of bit rates) is a choice that each developer or producer needs to make when burning a title to recordable media. "Some of the recordable formats," Ben explained, "aren't very compatible with the players out there. Probably the most compatible format at this stage is DVD-R for Authoring—the original DVD recordable format. DVD-R for General is the next most compatible, followed by the RWs and the pluses and minuses—things get a little tricky. If I'm going to be burning a disc for DVD delivery that I want to work on a wide variety of things, I'm only going to use DVD-R. The other formats are fine if I know the other person has a compatible player, but if I am just going to send it out to the world, I'm going to use DVD-R on high-quality media and I'm going to keep my peak data rate to about 6 megabits."

The loss of quality from the lower bit rate won't be severe, but it is a factor. The more noticeable problems from encoding at a lower rate typically turn up in the more challenging parts of the video. "When you're opting for DVD-R distribution," Ben said, "it is going to lower the quality you can achieve by a fair amount, but if you have good source video and a good encoder, it is going to look fine. It won't, however, look as good as a primo Hollywood movie. Depending on what the content is like, there will be a quality loss only when the video is at the peak, but the average is still going to be the same. The hardest parts of the video are not going to look as good, but the average, standard parts are going to look just fine."

Recognizing Artifacts

Video compression, even at its best, produces certain artifacts that can be discernible during decompression and playback. These artifacts might be barely noticeable, or they can be a disturbing annoyance that detracts from the overall viewing experience. Many compression algorithms try to remove video information that is below the perceptual threshold for the human eye—how well this succeeds can depend on both the degree of compression that is performed and the quality and integrity of the compression tools.

Getting the Best Video Compression Results

Figuring out the best techniques to use while encoding video typically requires long hours trying out different options, viewing the results, tweaking the parameters slightly, and then retrying the process. Ben has some thoughts, based on his long experience working with compression tools on both Mac and Windows platforms, for achieving the best video compression results.

"If you are doing any kind of long-form content," Ben said, "having a good two-pass VBR encoder really helps out a lot. You are able to distribute the bits in the file proportional to their complexity. The encoder will raise the bit rates for the hard parts and lower them for the easy parts. This can make a huge difference in the quality."

Professional-level work generally requires more-expensive, higher caliber tools, although the dividing lines between professional and prosumer are gradually being reduced with technological improvements. "If you are doing a professional DVD," Ben said, "you are taking either a film source or a video source that is shot well and then using a high-end compression tool. One feature is very important if you're working with content originally shot on film and transferred to DVD, such as movies, TV commercials, and so on. You want to be sure that you use tools and techniques that get the telecine process correct. That basically flags the frames in the output so that a progressive DVD player can play the frames correctly on a progressive DVD player. This also looks much better when the DVD is played back on the computer."

As for video encoder recommendations, Ben works with many different ones and tends to favor different encoders for different types of projects. "One of my favorite encoders," Ben said, "is ProCoder from Canopus. The new Discreet Cleaner XL has a pretty good MPEG encoder for DVD work, but it isn't very flexible for other kinds of projects. Heuris MPEG Power Professional is also an encoder that I like a lot."

Software Encoders Versus Hardware Encoders

Rapid increases in processor performance on both the Macintosh and PC platforms have brought video compression within the reach of software encoders. Previous generations of computing equipment couldn't compete with the raw, specialized performance of hardware video encoders, and compromises made to reduce the encoding times generally resulted in inferior quality video. Developers today have a choice: both hardware and software encoders can perform video encoding tasks very effectively. From a standpoint of encoding time, hardware encoders still have the edge, but some professionals, including Ben, still turn to software encoding to get the degree of quality necessary for certain high-end applications.

"Fast computers are doing a much better job at encoding," Ben said. A hardware encoder has to do everything that it is going to do with a given frame in about 1/30th of a second. A software encoder can increase or decrease the amount of time that it spends on each frame according to the level of difficulty."

"Basically," Ben continued, "you can vary how much quality you can achieve in the time that you specify. If you only have 20 minutes of content on a DVD, you don't need to sweat. The more bits you throw around, the less you need to use a slow, high-quality encoder. To get a 10 percent lower data rate, you might have to increase the compression time by a factor of 10—the ratio is pretty extreme. Small gains have a big cost in terms of the processing involved."

The end result, however, is paramount. "If you're going to produce a DVD that will be purchased by 10,000 people," Ben said, "this extra encoding time can be well worth the effort. ProCoder in its highest quality mode can work at about 15 times real time. That is at a totally extreme level where every possible sacrifice of time for a slight potential increase in speed is used. That is a very atypical setting. If you use the default for the encoder, it is closer to five times the real time value. It is a situation where you can spend a lot of rendering time trying to get slightly better playback quality. At times, that is an appropriate trade-off. Other times, it is not so appropriate."

Is there an interim time to shoot for that gives reasonable quality without overly increasing the encoding time? "If you have a good, fast computer," Ben said, "it is almost linear with the speed of the computer. A dual, Intel Pentium 4 processor-based machine using Hyper-Threading Technology is going to be much, much faster than a machine powered by a single Intel Pentium III processor."

"Most of these encoders," Ben continued, "are multi-processor savvy. I would expect that you could typically do a good job in approximately five times real time on the fastest machine. You can get a really decked-out machine for around three grand these days.If you're doing much video compression work, trying to limp along with hardware that is more than a year old really isn't worth the effort."

Embracing Interactivity

"When you are doing DVD," Ben said, "embrace interactivity. You want to get chapter markers in your project. You want to include good navigation. You want to do whatever you need to do to help the viewer watch what they want to watch. That will increase the value of any DVD title."

"You don't really need to do really fancy motion menus," Ben said, "but you really want to take advantage of the interactivity of the DVD format so that users can have those chapters available."

"The fundamental measure of quality for everything that you do is fitness of use. Don't focus exclusively on less important factors, such as: does the video look perfect? Focus, really focus, on the fitness for use. Have you designed the project so that your audience is going to be able to get as much value out of it as possible? Having the video look good, of course, is an important part of that. Having quality sound is an important part, as well. Having it be compatible with the DVD player that the user is putting the disc in is even more important. It is better to have mediocre quality that works than perfect quality that no one sees."

Inside a Criterion Collection Project

David Phillips, the Manager of DVD Development for the Criterion Collection, worked closely with Ben Waggoner on some of the difficult areas of video compression in the films of Stan Brakhage. Many of Brakhage's works included long sequences of hand-painted 35mm frames in which the frame content can change very abruptly—a particular challenge when performing MPEG-2 compression where averaging and interpolating frame content helps the efficiency of the compression process.

Hardware Versus Software Encoding

The Stan Brakhage title recently released on DVD by the Criterion Collection includes 26 films of the legendary experimental filmmaker broken down into two categories: photographic films and animation films.

"The encoding of the photographic films was handled quite successfully by the high-end Sony Vizaro hardware encoder that we use," David explained. "Since the overall length of the DVD video content allowed plenty of disc space, Criterion could use a very high bit rate for the encoding. Conversely, for most of the animation films, every frame is completely different. The hardware encoder wasn't really able to keep up with changing every single frame. A software encoder can take that process offline, so we are not trying to do it in real-time. The software encoder can also take as long as it needs to figure out the optimal way to encode. Most of the time in a high-volume production environment, software encoders are not an option because of the volume of much material. With software encoding, you have a whole extra step of actually digitizing the footage, Even if your encoder is very fast, even if it does near real-time encoding, that extra step takes more time. Since we had the luxury of time on the Brakhage project and since visual quality was paramount, we were able to take the extra time that software encoding requires."

Progressive Encoding for Frame-by-Frame Fidelity

Students of cinema often like to study the film by moving frame by frame through a scene, a practice which is compromised if the content is an interlaced 60i presentation. This problem can be circumvented by presenting the film content that has been transferred to video at its original 24 fps rate using progressive scanning techniques.

"The Sony Vizaro is generally very good at encoding film source material from videotape," David said. "The Sony can detect the 3:2 pulldown and then remove that as you encode the stream. For us, there are advantages to encoding our material as film. From 24p source material, there are 24 progressive frames that are encoded in the stream. That means that when people are stepping through frame by frame, they are not going to see any jitter frames caused by the extra fields added with the 3:2 pulldown. Many of our titles—and especially the Stan Brakhage piece—are viewed by film scholars and film students who are watching the films frame by frame to study the composition, the lighting, and other elements of the image."

"Any interlaced frames," David continued, "would be visually distracting and unfaithful to the source. One of the things that we are really excited about with the Brakhage set is that, for the first time, people would be able to step through Stan's films frame by frame. We recognize the inherent compromise in watching his films on video. There is no substitute for watching his films projected as films, but this is the next best thing that we can offer—a high quality DVD representation of his films."

What do you lose when you transfer a film to DVD? Clearly, each media has its own unique properties and character. David describes the difference in these terms, "There is a quality of watching a projected film, a depth, a dynamic range that you get with film that is not present in video. Film is a whole different medium.

Interviews on Disc

There are several interviews on the disc with Stan Brakhage where he talks about making his animated films. The films were so labor intensive that he was only able to complete about a second and a half of film per day.

"Being able to appreciate them on a frame by frame level," David said, "is something that we were really excited about when we decided to put out this collection. For the first time, people are going to be able to see the individual frames that make up his animated films. While he did step print some of them and they might be eight or twelve frames per second, many of them are actually 24 frames per second. This process has a quality all its own, but you certainly aren't able to actually see the detail in the individual frames while a film is running. So, it is very exciting to make this feature available to people. We feel as though this set was basically democratizing access to his films. Unless you live in a city with a Cinematheque, you probably don't have the opportunity to see his films. He is such an important figure in the film art of the twentieth century that it is a shame that his film work is not more easily seen."

Sourcing Original Materials

David points out that one of the guidelines that Ben Waggoner follows, which also dovetails with the work of the Criterion Collection, is starting encoding from the highest quality master possible. "Ben's basic rule of thumb is," David said, "that as far as encoding goes, 80 to 90 percent of the work of making digital video should be focused on making your source—your master—look as good as it possibly can. Then, the final five or ten percent comes into play getting the right settings on your encoder. Maybe there is one or two percent in there where it is certain mastery of your tools and knowing what to do if you find something wrong. So, for the Criterion Collection, the most important thing to begin the whole process is to produce the highest quality master that we can. Which begins with sourcing out the best possible original materials that we can."

This task is made more challenging by the fact that early prints can often be of quite poor quality or damaged in certain ways. "This really ends up being one of the primary challenges of releasing the films that we do," David said. "In films that are even twenty or thirty years old, the original negative will not have been stored properly or it will just be lost. So, what is the next best available material? A lot of times it will be an original inter-positive that was made. Sometimes there will be a fine-grained master that was made from the original negative. We essentially go down the line, looking at what is available. Oftentimes, once we find out if it is available, it is a matter of having the archive or the library either release the source material to us for transfer. In the case of a lot of the foreign films that we put out, we will arrange for it to go to a nearby transfer facility for a couple of weeks, since they don't want the element to leave the country where it is stored. Then we have to schedule a trip for a technician to go over and supervise the transfer."

Digital Restoration

"Previously," David said, "if we had an old film that had a lot of dirt and scratch problems, a lot of damage, sometimes we would try to do a wet gate transfer. Often, the people who own the materials have restrictions on what they allow you to do to it. A lot of the restoration previously had to happen after we had already transferred it to video and then were bringing it into our inhouse restoration department. We use systems from Mathematical Technologies (MTI), a company based in Providence, R.I. They produce a package which is now called *Correct.* So, that is one high-end dedicated film restoration program that we run. We use it for almost all of our releases."

"Digital Vision has a real-time hardware product called *DVNR,*" David continued, "which is their noise reduction and dirt and scratch removal system. It is a real-time hardware processor so we usually do a light pass through the DVNR system to get out the gross dirt and scratch artifacts, and then we will go in with the MTI system, *Correct,* and do manual fixes. We have four or five operators who are working from eight in the morning till eleven at night on this process."

Severe Damage

Restoration work, especially on some of the older films, can be a demanding task. "There is one film, *L'Avventura*," David said, "where the best master that we could find had some severe damage—dirt and scratches. One sequence took us two weeks to clean up maybe thirty seconds of film. The sequence was shot out the back of a motorboat, so there is water and waves and flashing going by, a lot of fast motion. All of the detail and the fast motion in the scene makes it very difficult for the equipment to separate the artifact from the natural image. That sequence was one that required detailed manual fixes—hundreds of fixes per frame, a very time-intensive process."

Supervising Transfers

After finding the best original source materials that are available, Criterion has a team of several people who supervise the transfers. "In order to maintain consistent quality," David said, "having someone supervise the transfer can make a huge difference. The technician or engineer at the transfer facility might have objective standards of what they are trying to achieve, but having somebody who is really experienced and bringing a certain subjective eye to it makes a difference. Ensuring that things match, that the color timing matches from scene to scene. That the overall density of the image is set properly and that the blacks and the whites represent the original film source."

During the transfer process, changes and enhancements to the image quality are often made scene by scene, shot by shot. The time code provides the basis for determining where corrections are made. "Once the film is transferred into the system, they can go back scene by scene and adjust the properties," David said. "It gets down to doing the final color correction at that level of detail to maintain a consistent source, a consistent image all the way through. At the same time, we are trying to overcome deficiencies in the original source material. Maybe the only thing that existed was a dupe shot that was inserted into the master, so it doesn't match in terms of the color timing or the focus or anything. For us, having that real attention to detail and transferring the films for our tape masters is really an essential step in the process. Having the best possible source ripples down. It makes the encoding that much easier and makes the restoration that much easier. It is a very important part on the front end. It is almost akin to if you are shooting a video, having the best possible lighting and exposure on the video so that you weren't having to fix it later on."

Mastering on D5

Restoration work and image enhancement currently take place using masters stored on D5, Panasonic's high-definition video format. The current workflow includes doing all the ongoing restoration work in high definition.

Origins of the Stan Brakhage Project

Peter Becker, the president of the Criterion Collection, has had ongoing involvement with the Stan Brakhage release dating back to the days when it was first considered as a Laserdisc title. The economies of scale didn't fit the model of an expensive Laserdisc at the time, but the rise of DVD technology, with its less expensive manufacturing process and wider base of installed players, brought new life to the project.

The transfer of Stan's films to DVD from film was not a decision that was made lightly. "We talked about it with Stan," Peter said, "and we talked about it with Marilyn (Stan's widow who is in charge of Stan's estate and has the heavy burden of watching after his films). One of her gravest concerns is that DVD not supplant the films, but supplement the films. It is one thing for students to have access to the films or for a broader commercial audience have access to the films than ever before, but if the result of this is that the films will no longer be projected and people will no longer rent the prints, that is a disaster. It is a little like replacing all of the great works of art in the Louvre with postcards and posters of the works of art in the Louvre. You have to remember that what you are looking at when you look at the DVD of Stan Brakhage's films is a reproduction. The artwork itself is originally a piece of printed film. It needs to be shown projected on a wall or on a screen. There is a fundamental difference between reflected light, which is what we are looking at when we look at a projected film, and projected light, which is what we are looking at when we look at a backlit screen, a video screen."

Stan Brakhage was alive when Criterion started the project in 1996, but shortly thereafter he became ill. The anticipated Laserdisc market was dwindling and Criterion was concerned that transferring the films to video in a quality telecine suite and then mastering these discs for an estimated two million players would not be a wise financial move.

"There was a good chance," Peter said, "that if we had put out a set of Brakhage discs, we would have sold just a few hundred copies: maybe 500 or 700 copies. Who knows. Once DVD really settled in and we had a sense of what the size of the marketplace was, it became feasible for us not only to make DVDs of them, because the DVDs themselves were cheaper to manufacture, but actually to manufacture the new film elements from which to make the DVD. So we went back to the negatives and we made new fine-grains and interpositives from the original, from Stan's original negatives at Western Cine, which was his lab and long-time collaborators. They printed everything for him."

Working with Western Cine

The Western Cine lab that had been such a central part of Stan Brakhage's work belonged to a well-respected man named John Newell, who died in 2002. "He was the recipient of the Anthology Archives Film Preservation honors the year before he died," Peter said, "and I think it meant a great deal to him. He got up and gave a speech that night and said, 'I have never been given an award in my life.' He had been helping students and independent filmmakers for years and years. Among those people, most prominent among them, was Stan Brakhage, who printed just about everything he ever did at Western Cine. At various times, there were different people at Western Cine that Stan would work closely with on optical printing techniques and such."

Black Ice

"There is a film called *Black Ice* that I think represents a really interesting challenge and it shows Stan's work well. It is inspired by a fall that Stan took on black ice on a driveway and it led to temporary blindness. Stan had all kinds of vision problems. A lot of his work is really about vision. What is interesting about this piece is that it was very difficult to compress. It is quite beautiful, a hand painted film that is also optically printed. It gives you an incredible sense of depth, even though you are looking at these two-dimensional film frames that have been hand-painted. Because there are these hand-painted two-dimensional film frames that have been optically printed to create an incredible sense of space. That is a great example of Stan as a collaborative artist, doing this very private work of painting on film frame by frame by frame and then taking it to the lab and collaborating with the folks at Western Cine. I think there is even a credit on that film that talks about the importance about how the film should be seen as a collaboration."

Black Ice, which is about five minutes long, appears on the DVD bundled with this book.

Stan's Vision

Throughout his life, Stan Brakhage had to contend with a number of vision problems. These problems never caused him to stop working. "Stan's life and his work were very deeply intertwined," Peter said. "Whatever was going on in his life became the subject of his work, although not necessarily explicitly. And, his work became the substance of his life. When he had his first child, the documentation of that first experience of childbirth on film was a very unusual thing to do in 1956. It was terribly ground breaking to film that experience. Stan explained that it was the only way that he could get through that experience. The camera grounded him."

In the Morgue

Another long film on the Brakhage disc is called *The Act of Seeing with One's Own Eyes,* which was made over the course of a few weeks. The time was spent in the morgue in Pittsburgh where Stan was essentially filming autopsy procedures. Autopsy comes from the Greek root, the act of seeing with one's own eyes, seeing for yourself. Stan was given this opportunity as part of a trilogy. One film he did while going around with the police and the other was a hospital emergency room. The other was at this morgue. "To hear Stan tell it," Peter said, "he is hanging onto that camera for dear life. He could not possibly have inhabited that room for two weeks if he hadn't had the camera as his motive and method for being there."

Saving the Negatives

Sometime during 2001, Peter Becker got a call from Bob Harris, a well-known film restorer. Bob announced, "Western Cine is going to close any minute. John Newell is going to retire. If we want to do this Brakhage project, now is the time. Can we start ordering interpositives and fine-grains?" The Criterion Collection immediately got the process started and the resulting interpositives and fine-grains have now been deposited at the Academy of Motion Picture Arts and Sciences where the most modern preservation techniques will ensure a long lifespan.

"This was a case," Peter said, "where the DVD project has had an enormous effect on making Stan's work accessible to a new generation of viewers who never even knew that he existed before. We get all kinds of letters about it. Lots of people have learned about his work from the fact that it is available on DVD. But another important sideline of this project is that the films themselves, those high-quality, original pre-print preservation elements created in order to make the DVD, are now being stored in as good an archival circumstance as you could ever imagine. The Academy is really state of the art. For us, it is a win-win. We are very proud to have made the films available. We are very proud to have done something to help preserve the films. And, now, I also just feel some obligation to help make sure that the fact that the DVD is out there does not keep the films from being projected."

Stan did a very good job in the later years of his life to make sure that his work was going to be preserved on film. He chose very carefully, I think Stan trusted Criterion because he knew what we stand for and he knew our approach: the reasons we originally set out to get director approval of content. The reason we originally set out to present films in their original aspect ratios or to invite directors into telecine sessions or to do all these other things that are considered unnecessary and expensive. When we first started doing them in the middle 80's, I think Stan really recognized that the motivating force behind this was that we really did care. We didn't want to see the films end up either lost or mistreated. Whatever was in keeping with his wishes, that is what we were going to do. And we did. Stan was the person who decided that Fred Camper should be his eyes in the telecine room. He said, 'Fred has watched these films projected more than I have at this point. And he sees better than I do, so you should use him for approving the telecine masters.' If you've been to Fred's Web site, Fred is an outspoken anti-video man or, he was, at least until this project started. If you go to his Web site, there is this piece that he wrote back in 1986 all about how video is horrible."

Fred sat quietly through much of the telecine process and then after about 45 minutes, he said quietly, "You know, this is quite good. If this is what video has become, I may have to reassess." Fred ended up supervising the telecine process and helping assure the quality of the resulting video on the Brakhage DVD.

Variations

"Stan himself believed," Peter said, "that the printing process, as fallible as it is, was part of the making of the artwork and that there wasn't a right color for a print or a wrong color for a print. When a print came off a little bit too yellow or a little bit too red or a little bit too green, Stan saw that as some of the interesting variation that occurs when you are making mechanically reproduced, chemically processed art. That every print is unique. And that variations within reason are to be celebrated."

Stan was alive through the whole film-to-video transfer process. He was able to see most of the finished work, except for some final cuts of video interviews that had been acquired from a friend of his. The video dated back to the 90's with some more recent clips in 2000.

Progressing from Laserdisc to DVD

Criterion's original specialty was making Laserdiscs and many of the current features that appear as DVD titles were originally developed using the capacities of Laserdisc to show still frames or play multiple audio tracks.

Laserdisc, however, proved to be a very modest marketplace. "I don't think there were ever more than about 2 million players in American homes," Peter said. "DVD exceeded that number very rapidly."

During the Laserdisc era, the Criterion Collection won over many fans who appreciated the care and attention that went into the release of both classic and little-known film works. Surprisingly, many of those early customers became early adopters of DVDs. "We did immediately have a presence in the DVD marketplace, as soon as we entered, that we wouldn't have had if we hadn't been laboring away all those years on Laserdisc. On the other hand, it is no surprise to us that now that we finally have a market that is broader, that is a mass market, we are finding more customers. To begin with, there are simply more players out there. Also, many of the films we are working on are huge classics in world cinema— *Rules of the Game, La Strada, Tokyo Story,* and all these great international classics. Then, we have *Straw Dogs* and *The Last Temptation of Christ* and a lot of the Hollywood stuff that we have been doing. It doesn't surprise me at all that we are finding a broader audience. I am very gratified that a lot of the work that we have been doing that started out on Laserdisc has been broadly adopted on DVD."

More information about the recent releases of the Criterion Collection can be found at: *www.criterioncollection.com*. More details about Ben Waggoner's compression techniques can be found at: *www.benwaggoner.com*.

12

Case Study: Supporting Independent Filmmaking

Edgewood Studios confounds conventional lore about movie production. Located in the most rural state (Vermont) in a city that is by no means a major metropolis (Rutland), Edgewood Studios serves as an example that you can follow your dreams, make a good living, and give something back to the community along the way. Co-founders David Giancola and Peter Beckwith did just that, turning their love for movies into a sustainable business. From modest beginnings doing the wedding circuit and creating industrial videos, Edgewood Studios now handles million-dollar plus feature films, facility and equipment rentals for production groups working in Vermont, and a variety of independent features, many of which serve as training vehicles for the next generation of filmmakers. They're actively involved with distribution of video features internationally, production of their own feature material, and work-for-hire projects destined for cable television broadcast.

The lessons presented in this chapter, and the underlying advice offered by Dave Giancola, veer off on a somewhat different course from the rest of this book. Dave knows and trusts film and thinks it is far too early to join the stampede to digital video, particularly when the standards continue to change at a blistering pace, rendering today's state-of-the-art cameras and editing tools into tomorrow's doorstops. Some of the 16mm and 35mm cameras in his rental collection date back more than a decade, but they can still produce high-quality films with the rich colors and depth and contrast that characterize film, as opposed to videotape. Those films can serve as masters to support the next generation video formats. When high-definition DVD standards are in place, Edgewood can return to the original 35mm negatives and digitize the material to fit the new standards. From one high-quality master, any number of future potential digital video formats can be supported simply by converting the film content by means of the telecine process.

Edgewood has been enjoying a resurgence of interest in their earlier work, which is now being re-released on DVD and generating new sales in the distribution channels.

This chapter offers a perspective on the importance of distribution for filmmakers, the techniques for producing films under low-budget conditions, and the increasing importance of DVD as a medium to reach new audiences and grow a business.

From High School Student to Moviemaker

If the thought of several years of film school to become a moviemaker seems daunting, you might approach the problem the way that Dave did. Just jump right in and start your own video business. As Dave explains it, "Edgewood Studios started in 1987. I was literally just out of high school. I really wanted to get into the movie business, but I was not happy about the prospect of four years of college or going to film school and playing by somebody else's rules. Pretty arrogant. I just wanted to get started. I had some friends who were already going to film school and they were stuck in the theory, barely touching the cameras. I decided: I am going to start my own company and get going. So, I forced my way into some office space, got a bank loan, got some video equipment, and started shooting whatever anybody would pay me for. Which was primarily, at that time, weddings, legal depositions, and even a funeral or two."

While the business gained momentum, Dave was also working on a variety of creative concepts and developing movie ideas. While drinking in a local bar, he met Peter Beckwith, a Cornell graduate who was managing the bar and also cruising around Vermont enjoying the role of ski bum. The two began talking and discovered that they had a similar outlook on business.

"I was very passionate about making film," Dave said, recalling the initial conversation, "and he was very passionate about business. There was a really good mix between us." Shortly afterwards, the sole proprietorship became a partnership.

"We actually were doing really well with Edgewood as a video production company," Dave recalls. "We stepped out of weddings after a couple of years. It was great training for documentary work, but we really had had enough of it. I knew that I had booked my last wedding when I got in trouble because I didn't videotape the hors d'oeuvres at a wedding. The mother of the bride was furious at me because I didn't videotape them and suddenly they were gone. That's when I said, 'OK, I'm done with weddings now.'"

The Commercial Route to Film Making

From this early experience with video production, Edgewood Studios began producing television commercials. "We started doing regional cable commercials and those turned into some New England based commercials followed by a few national commercials. Although we were doing pretty good business with the commercial work, our passions led us to dramatic work. We took some of our money and made a short film called *Ten Minutes*, which was based on Will Eisner's comic book story. It did really well, won some awards, and got into some festivals. It was a great start."

Turning to Film

At this point, Edgewood Studios turned to 16mm film for their production work, a reversal of the typical progression from film to video. "Like a lot of people at the start of my generation, the usual progression was that you started on film and then went into video. We started in video and discovered film along the way and really fell in love with the image."

Dave made a conscious decision at that point to get started making movies or to get out of the business completely. The first movie production was launched not much later. "We put together a crew and $40,000 to make a film called *Time Chasers*. It was shot, literally, on weekends over a six-month period on 16mm film. Everybody who worked on the project was deferred. The budget, which started out at $40,000, ended up being $150,000 when we were finished two or three years later. The film was commercial enough that we found a distributor and it really got us started."

With this first successful entry into the low-budget film business, Edgewood Studios started work on a second film, *Diamond Run*, which garnered even more success. The next film out of the chute was *Pressure Point*, which had about three times the budget of the prior film. "Before we knew it," Dave said, "we were making films with million-dollar budgets.

The quality of our projects got better each time. Somewhere along the line, we met with partners who wanted us to do a multi-picture deal and we started building a studio. The next thing I know, I'm sitting inside a 50,000 square-foot building thinking, 'OK, I own a studio now.' That is how we got to where we are today."

Business Philosophy and Direction

The evolution of Edgewood Studios was guided by a shared business philosophy that Dave and Peter developed. To be closer to his family and to benefit the company, Peter relocated to California, settling in the heart of Los Angeles in Burbank. "Peter looks out for our financial and creative interests out there," Dave said, "including those interests that are tied to distribution."

Nonetheless, Dave feels grateful that the Edgewood studio is located in Rutland, Vermont for the overall quality of life that the region offers. The satellite office that Peter operates in North Hollywood—because of its proximity to so many key industry contacts—makes it possible to work actively within the film business. "You really need to have a least a toe in that pond," Dave said. "Edgewood does a number of things. We co-finance low-budget features with other partners, building a library of our own films. We get this library of films distributed worldwide. We also do work-for-hire projects, such as the Porchlight Entertainment movies we have been doing for Pax Television. Because of our expertise in putting a lot of production value on the screen, we get hired to make certain kinds of movies. We have become the experts of the mini-disaster movie, as you may have seen from our Web site."

Edgewood Studios also represents other producers, working with local, national, and international filmmakers to handle their work in the marketplace. "The waters of distribution are filled with sharks," Dave warns, "who are perfectly willing to take your film and not give you any money. We are a credible source. If we feel that a film has a life, we will take it for a small percentage and help the filmmaker out. Sometimes we come in as film fixers, where a filmmaker comes to us with a film that is not complete. Maybe the filmmaker got the film shot, but then couldn't afford to get it edited or mixed. So, often we come in after principal photography has been finished and partner with the filmmaker. Because we own the facility, we can also partner with the filmmaker from pre-production on. Sometimes we do it for fees. Sometimes we do it for deferred fees. Sometimes we do it partially for fees and a piece of the film. Sometimes we do it for a percentage of the film. There is no single way that we work, but we look to partner with filmmakers who are talented and have a good work ethic and understand the marketplace a little bit. Those are the filmmakers who are generally a good fit for us."

This essential but non-glamorous work extends Edgewood Studios' business model into areas that require a healthy understanding of the available distribution channels, the changing trends in market conditions, the tastes and preferences of worldwide audiences, and the going rates for different types of feature films. As Dave explains it, "Everybody gets excited when you have the money to make a film, but nobody wants to talk to you when you are out of money and trying to take the film around and find a distributor."

Nurturing the Next Generation of Filmmakers

Part of the Edgewood Studios mission involves nurturing the new generation of upcoming filmmakers, setting up channels so that they can get their work done, providing equipment and support, actively working to manage the distribution of the final work.

This approach offers a host of side benefits. "Not only is that good for the filmmaker, it is good for us because those are the filmmakers that get involved in the productions that we own and co-finance with our partners. This is a great way to find talented filmmakers, to look at what they've done in the past, and to build a relationship."

The early experience dealing with international distributors has made a big difference in Edgewood's approach to marketing. "Our films really started with success internationally, before we found success domestically. The international marketplace for television and videocassette was much stronger when we started in the mid-90's. That was the place for independent film to find an audience. But, recently, with the introduction of DVD, the situation has gotten better. The cost per piece is down. Distributors are more willing to take risks."

The equipment rental side of the business, Dave has found, leads to many long-term relationships. "We offer low-cost rentals to filmmakers all over the east coast," Dave said. Sometimes we partially defer the costs and this builds a relationship. The approach works out really well. We are really interested in building relationships, as opposed to having a one-film hit. The filmmakers who last and make work that resonates—financially successful films—are the people who have been around. Filmmaking is a craft that you have to learn. Like any other profession, the people who have the experience and the dedication, the people who stay around longer, are the ones who are successful and whose work you want to see."

Approaching DV Cautiously

Dave approaches the exuberance over the popularity of DV with a cautious attitude grounded in the hard realities of the marketplace. He also admits to a healthy skepticism about the life span of the format, given the notoriously unpredictable nature of the electronics industry.

Dave said, "I have to preface my skepticism with this: I think that DV is a spectacular medium for people to learn, to experiment with, and to take chances. And, I think there are great opportunities within this medium. Having said that and having been in the video business where the format keeps changing every couple of years, so you've got to keep upgrading, it is very clear to me that the next five generations of technologies are sitting at the top floor of some offices in Japan somewhere. And they plan to release them when they are good and ready."

In Dave's estimation, the manufacturers of DV equipment have found a new, wide open marketplace within the independent filmmaker community. "I can't tell you how many times someone has come in and said to me, we're going to shoot on DV—I just bought a camera for three grand. What they don't know is that I go to the markets, I go to Milan, I go to Cannes. DV is not delivering consistently or well internationally. You might be able to do OK with your DV film on a domestic videocassette or on DVD, which is more forgiving. But, for domestic television, for international television, for international videocassette, where you have got the format changes for PAL, there are increased expectations. They want the Hollywood look."

Pausing reflectively, Dave continues, "There are just not the opportunities that people think there are. I know a lot of distributors who will reject a film the minute they discover it is DV. I was at a distribution company two weeks ago in L.A., looking for good material, and they said, 'Here is a pile of DV tapes in the corner.' The corner of the room was full of DV tapes, screeners of films that were shot on DV. I am concerned that there is a lot of hype going around right now. That hype directly drives large electronics manufacturers who are making a lot of money."

Edgewood Studios offers a PAL BetaCam camera for rental, but that's about the only concession they currently make to the realm of video. It's not so much being against the format as being against the hype. "I happen to think that DV is a great opportunity and I think that for the right project, there is a lot that can be done," Dave said. "But I think you need to look at your market before you buy all the hype about DV. Be very critical of it. Be able to separate the fluff from the facts. A lot of people that I talk to who are completely pro-DV can't really explain the differences between DV and HD. This is a concern, because they are totally different things. I can still buy a BetaCam SP camera that has been around for 10 years and can still give you a better picture for the same or less money. So, the whole DV technological breakthrough is really a financial breakthrough, in terms of the electronics companies trying to open up a new marketplace."

Fighting the Stacked Deck

Promising developments, such as the adoption of digital cinema systems in independent theaters, still don't eliminate the barriers faced by the independent filmmaker. As Dave said, "The major studios have theater chains, including Landmark Theaters, locked up forever. That is not going to change. There is still going to be competition for screentime because that theatrical screen time, whether profitable or not, at the very least drives domestic videocassette distribution. I've had filmmaker friends who released 35mm prints to the theaters, films that were outperforming every other screen in a five-plex. And, the film got bumped because the studio had another release that they wanted in the theater. Theater owners rely on the major studios for a consistent flow of product. There is really no leverage for the independent filmmaker. In this instance, I don't think technology is going to change everything. There is still going to be competition for those screens and those bodies in the seats,. The technology doesn't change the real estate or the popcorn or the overhead or the labor."

With all this said, Dave still thinks there is a place for DV in the film world. "I think the opportunity is in the world of festivals, if you are willing to be patient and take your time with DV. If you are at the point as a filmmaker where you only have five grand in your pocket and you want to make a film and you want the flexibility of DV, I think it is fine. Ask yourself, however, what is the marketplace for this film? If it is a little horror movie and your budget is small enough, you might be able to make it. The opportunities exist in DVD now because you've got video distribution companies that are more open to content shot on DV. That is encouraging, but it is only one small venue for your film."

"I know some independent filmmakers," Dave said, "who are doing OK through the domestic pipeline with videos and DVDs. If you are doing something that is really about the actors and the script and the story, you can run the festival circuit and see if you can make it happen. Keep in mind that the odds are maybe not as good as taking your money to Vegas. But, in terms of thinking that your DV film is going to have the same kind of attention paid to it as something shot on 16mm or 35mm, right now it is just not happening."

Shooting Film and Transferring to DVD

Dave feels that the way to deal with the rapidly evolving video standards and changing equipment is to use film as the production and archival medium. That archived film negative then serves as the source for any number of video formats through a relatively simple (though not inexpensive) conversion process.

"My feeling," Dave said, "is that you have to look at technology for where it helps you and recognize where it does not. It is very hard if you are just starting out to get a really good sense of what is going on in the industry. Where are the real technology benefits to the independent filmmaker? The fact of the matter is, we are in the business of making movies. The smaller stuff we do—you probably saw *Arachnia* or *Moving Targets*—are really low budget. If there was a way I could shoot on video and get away with it, I would. But, if I am going to spend the amount of money that you would spend buying a house, as a business owner, I am going to want an asset that is going to be around for a long period of time and is going to have the widest possible potential to be distributed."

If I shoot on film right now," Dave said, "I know that film is very stable archivally. I also know that whatever format comes up, I am going to be able to telecine to it. As the major electronics manufacturers start to put out HD and then SuperHD and then double SuperHD, you have a good high-resolution film negative that you return to. That negative holds up archivally. When your film budget tops $100,000 or $150,000, with the amount of money you are spending on labor and everything else, even though your laboratory and film costs are a big expense, the negative secures your investment. This is my perspective based on the way that our company makes money."

Figure 12 - 1 **Car-mounted camera**

Weathering Format Changes

Given equipment costs and the expense of upgrades, Dave expresses a very clear concern over the rapid progression of formats and the expectations of the marketplace. "Even on our very small films," Dave said, "our buyers are demanding 5.1 mixes. I don't have enough money to really do a good mix on the film as it is. But, now they want a 5.1 mix without paying me any more. I keep watching technology like a hawk, reading everything and waiting for that sweet spot to do the upgrades. Sometimes it is tough to find that sweet spot."

"For some of our older films," Dave said, "we have found that they enjoy another life on DVD. It is amazing that a format can do that, but it does. DVD is cheaper to produce—therefore, you can take more risks on it. There are titles that we released on VHS a long time ago, that might not have gotten re-released any time soon. But, because there is a new format that is collector savvy, the film may have a kind of cult attention. People want to take a chance on it. It is fresher. For our bottom line, that makes a difference, particularly if you can re-release a title that has already been paid for."

Producing DVD Extras

As part of the rising expectations of distribution companies and audiences, adding DVD extras to titles is rapidly becoming a de facto requirement. "I swear," Dave said, "within the span of about two years, DVD Extras went from becoming 'Well, if you've got em…throw em on' to 'Oh, you have to shoot a behind the scenes video. Can you get the cast back for this or that?' Nobody is paying any more for the movies. But, for whatever reason, DVD Extras has become standard. There is nothing better for a filmmaker than being able to talk about yourself for an hour and a half. Or, doing your own documentary promoting your film career. People watch these things and collect them. I think it is good for film collectors. I think it is good for filmmakers."

Edgewood Studios generally negotiates with a video distributor to provide the DVD rights and then submits the materials for the DVD Extras portion of the title. "We're involved in delivering a Digital BetaCam of the film, for example, and delivering a DAT tape of the audio commentary and a documentary."

To gain the additional material to round out a title and provide a deeper look at the production side, Dave sometimes goes through the outtakes to find promising segments, and, working from the early edit decision lists, decides which additional segments to telecine. This makes it possible to include alternate scenes and material that wasn't in the original movie.

When lab services are needed beyond the scope of what can be done inhouse, Dave generally relies on the efficiency and expertise of the larger labs. "Usually, when we are using lab services, we go to N.Y. or L.A. Generally, we send work to L.A. because L.A. labs, in general, have been cheaper for us, even including the cost of FedEx. Because we're located in Vermont, I've got to FedEx everything. I'm certainly not going to courier stuff to the lab. It is five hours away to anywhere. With FedEx, I can literally go for the best lab prices and quality anywhere."

Doing the Commentary

"When we first started, we did commentaries and documentaries all the time, just out of hubris. We thought that the stuff we were doing was so great. Then, as we went along making films, we became more jaded and we stopped doing those video behind-the-scenes things. I know on *Arachnia,* which is out now on DVD, I didn't have as much fun shooting the movie as I did sitting with the cast doing the commentary. It was just a riot—it was so much fun. We got about five cast members back to do it. Here's a tip for anybody doing DVD commentary. I had never done it before; this was the first commentary that we had done from scratch where I had a bunch of people in the room. They had not seen the film yet. I screened the film and everyone talked all through it. Then, we went back to record the commentary and everybody clammed up. If you listen to the DVD, I'm trying to stimulate the discussion and prodding them to talk. Everything changes once the microphone is turned on."

The commentary was synchronized to a time-coded DAT recording that could then be integrated into the DVD content during the authoring process. Dave was determined to keep the commentary light and fun. "I had listened to a bunch of commentaries and I found a lot of them to be pretty dry and serious. And, I had always wanted to make the style of the movie set the style of the commentary. I wanted to make it fun. The movie was fun—it is kind of tongue-in-cheek. On a movie of that budget, a lot of the people are there because they love film—they are not there just for a payday, which is refreshing. We went in and got a timecode DAT and we put it in our mix room so that we had the film on a projector screen. Acoustically, the room is really great. We miked up the room and sat everybody in the mix room and just ran the movie. We mixed the movie soundtrack down to a comfortable level where we though we could make it all work. The plan was just to run the movie and go with the audio without editing. On our budget, you can't. We just ran the movie and warned everybody that whatever they say is spontaneous. I think the commentary is hilarious, but then I was in it, so I don't know. It *was* a lot of fun and it was a great way to do it and people got a lot of insight into what was going on in the film."

Becoming a Filmmaker By Making Films

"I am a big believer," Dave said, "that you only become a better filmmaker by making films. Which means that you use the tools that you have on hand and work with the budget you have and then you just go make your film. I don't want anyone to think that I'm saying that if you can't make it on 16mm, it is not really a film. I don't think that is fair at all. That is just as inaccurate as saying that DV is equal to film. I think that the only way that I have ever advanced in my career is to keep working. You win some, you lose some. You make some good stuff. You make some bad stuff. But everything gets better with the perspective and the skills that you learn by going forward. DV is not a panacea for good production values or good work ethic or talent. And hard work. That is what I think the lesson should be for people. There is more to making a movie than just the technology involved and the camera. I think that technology is there for us to use. It doesn't necessarily create a wealth of opportunities, even though the advertisers may promote it that way."

Figure 12 - 2 **A bouyant platform for filming**

Contending with the Realities of Distribution

Dave believes that a good story can be told on any medium, but the realities of distribution are a critical concern. "It is very hard to get your work seen. It is very hard to have your work break out. Even if your work is good, there are a lot of political reasons in terms of what does or doesn't get into the Sundance Film Festival, for example, that don't have anything to do with the quality of your film. It is important to recognize that and understand it for what it is. So, you can succeed on your own terms. You can avoid debilitating disappointment (which I see a lot of). I don't think that is productive at all. There are filmmakers that work with us who live for one film. They get such disappointment and financial fallout from the process that they sometimes don't continue. There are things that can be learned from what we are doing at Edgewood that will hopefully open up people's eyes as to what is and what isn't going to work for them."

Rubbing Elbows with the Stars

Edgewood Studios hasn't yet worked on any 50-million dollar Hollywood blockbusters, but they've worked closely with a number of luminaries from television and the cinema world. "My new saying is," Dave said, laughing, "if you are working with Edgewood Studios, you are either on your way up or your way down. We have worked with Larry Linville (who has since passed away) on one of our films. He was Frank Burns of *Mash* fame. We have worked with the new hot little thing, Kate Bosworth, who was just in *Blue Crush* and she is now in *The John Holmes Story*. We are representing a film produced by the guys from Project Green Light, Kyle Rankin and Ephram Potelle, who we met because they're from Maine. There have been a lot of those stories where people are passing through our world."

"What is interesting about it," Dave continued, "you learn a lot about who these people really are based on how they treat other people. And, at the end of the day, that is really more important than their success. I know a lot of super-famous people who are really unhappy with themselves and the world around them. What is the point in doing that?"

Boosting Regional Filmmaking

Edgewood Studios works closely with the Vermont Film Commission to draw new filmmaking projects to the state, both from independent filmmakers and the major studios. "I think that the environment for regional filmmaking has gotten better in most states and in Vermont it has gotten a lot better," Dave said. "They have passed some legislation that, while not ground breaking, is certainly helping things. I think that among filmmakers there was a movement that started about 15 or 20 years ago where filmmakers left Hollywood and started to explore places like Texas and Florida. Regional filmmaking started to break out on its own. I think you are seeing part of that here."

"For a lot of the huge movies that came into the state," Dave said, "for better or for worse, the draw was the bucolic beauty of Vermont. In the movie *Cider House Rules,* everyone just loved the locations, the beauty of Vermont—they could make it work. They could also find a crew here, which was wonderful. The same thing with *What Lies Beneath,* the Jim Carey movie. They came here because I think they wanted to hang out in Vermont for the summer and just have a good time. And they did."

Business generated from movie production companies coming to Vermont to shoot helps keep the revenues flowing at Edgewood Studios as well. "When larger shows come in," Dave said, "we have a company called New England Lighting and Grip. We rent out lighting and grip for everything from the Budweiser guys doing a Clydesdale commercial to the big movies, where we get the overflow, to the little independent films. We help them all out. It is amazing to me the number of movie stars that have second homes in Vermont. We do a lot of looping for actors. They come in here and they can go to their place on the lake or in the mountains and then do looping on their movies—we make the arrangements with their studios to handle the details. An amazing amount of work gets done from our humble location in Rutland, Vermont. The longer we are around, the more people know about us. The work just keeps coming to us. I am still amazed myself that we are doing as well as we are."

More information on Edgewood Studios can be found at: *www.edgewoodstudios.com*.

13

Case Study: Creating a Workout DVD

Combined with relatively inexpensive DV media and camcorders, DVDs offer substantial opportunities for small businesses and individuals intent on launching an original production or a series of titles. After working closely with Heuris for several years as both a product manager and consultant, Jeanette DePatie found the lure of DVDs irresistible and set out to create her first original work: an exercise video that is unique in its graduated approach to daily workouts. Capitalizing on some rarely used features in DVD Studio Pro, Jeanette drew on her expertise in product development to design and produce the first title in what she hopes will be an ongoing series. *The Fat Chick Works Out* encourages women to drop society's body shape obsession and to work out for good health and the sheer fun of celebrating movement and life. The spirited workout sessions, planned after careful consultation with physicians knowledgeable about exercise physiology, are designed to gradually increase in intensity over several weeks.

As might be expected from someone with experience in product management, Jeanette planned every aspect of the DVD creation in meticulous detail. Rather than inhibiting spontaneity, she feels that this approach freed up everyone involved in the production to stop worrying about the myriad details of the project and to relax and enjoy themselves. Planning is particularly crucial to best use resources during a video shoot when every minute of wasted time adds expense to a project. Jeanette's guidelines for steering a DVD project through successful completion echo her own experiences in designing and executing the production.

Designing a Progressive Exercise Video

Developed from conception as a product for retail sales, Jeanette DePatie's approach to workouts follows a few basic assumptions. First, people getting started with exercising get discouraged if they have trouble keeping up. A workout should offer a challenge, but that challenge should be very gradually introduced. And, perhaps most importantly, a workout should be fun and filled with energy. Not many people enjoying plodding along to a military cadence led by an unsmiling, perfectly conditioned aerobics specialist. "I'm making a few small changes," Jeanette said, "but the is going to be available for retail sale pretty soon. It's called: *The Fat Chick Works Out*. And I'm the Fat Chick."

Jeanette took advantage of a little-used feature in DVD Studio Pro to introduce a unique advantage to here title. "This project," she said, "is the first exercise video that I know that is truly progressive. I used the Stories feature in DVD Studio Pro to create pointers so that each week the exercise portion of the video is longer. It progresses for 12 weeks. Each week, you go in and there is a little Peep talk, which is the rah-rah thing with your personal challenge. Then you go and learn your steps for that week. Each week you only learn one or two new steps, because this is really designed for beginners. You can always go back and relearn the steps, but Week 1 is maybe 15 minutes, and Week 12 is 45 or 50 minutes. It is organized the way that doctors really like to see beginners exercise."

The DVD, designed for playback on both set-top players and computer DVD drives, segments the content into a series of categories so the workout is guided by the selections. "The viewer goes in and picks a week. Let's say that I pick Week 6. I just hit the Exercise button and it automatically plays my warm-up and then it plays the exercise part through Week 6 and then it puts the Cool-down on the end. I did that all with the Story feature. It is a pretty simple idea, but nobody else seems to have done it."

The video shoots were handled with Canon GL2 miniDV camcorders from Donna Donahue's *Cameras Rolling* and some live switching. The editing of the project was accomplished in Apple Final Cut Pro. Before Jeanette shot so much as a minute of video, however, she had planned the project so that every aspect was budgeted, scheduled, and calculated, as described in the next section.

Planning: A Key Requisite

Jeanette believes very strongly that planning is one area where people starting out new DVD projects fail to put enough attention, whether they are filmmakers, marketing specialists, or developers. "Before I did any editing and before I did any programming, I built specifications and flowcharts," Jeanette said. "I knew exactly what buttons were going to be on what screens and exactly how many screens I was going to have. That level of planning allowed me to create a fairly sophisticated production with a very low budget, because it was all very efficient."

As Jeanette sees it, people typically get deeply into a project before they even work out the basic framework for their DVD presentation. A lot of grief can be eliminated by doing some prototyping and figuring out how all the parts are going to fit together ahead of time. "I know a lot of people call our tech support when their project collapses at the last moment. They get this idea for how they are going to make a DVD. They go out and shoot everything. They go out and edit everything. They put their DVD together. Then they start testing their concept. That is not really a good plan. If you get in the DVD-building stage and you find out that something you thought was going to work isn't going to work, you don't want to have to go back and reshoot around a new DVD design. You really need to test your DVD design—just use some dummy video and test your concepts. Am I going to be able to use subtitles this way? Am I going to be able to use the story feature this way? Actually set up the structure and then burn a disc."

Having worked closely with the tech support group at Heuris, Jeanette also knew that testing on different equipment during the development process can eliminate a host of future problems. Not having a test lab equipped with models of the latest DVD players, she took a rather innovative path to testing interim copies of the project. "I've gone to Circuit City and asked, 'Can I stick this disc in one of your players and see if it does anything.' And usually they say, 'Yeah.' Then I know that the idea is really going to work, as opposed to working in theory."

Part of the challenge is having a clear idea of what you can accomplish on the DVD. "Then," Jeanette said, "if you go to the people involved in the production, you can say, 'This is what I want to try to accomplish."

Budget realities drive many aspects of any project. "If I could have afforded it," Jeanette said, "I would have shot the whole thing in HD format, using a couple of Panasonic HD cameras, some professional camera people, and such. But I was paying for this project out of my pocket. Since it is my first production and I don't have a track record yet, I have got to just get something in the can."

Working from a Production Book

At the earliest stages of the project development, Jeanette created a production book. The two hundred pages covered everything from treatments to costume concepts to a complete script to a complete DVD specification with flowcharts. The production book also included tests that were run in preparation for DVD creation, market research, and other details that helped streamline activities when the project shifted from pre-production to production.

"I've been a product manager," Jeanette said, "so I think in those terms. I think in terms of what the overall project will look like. You don't go to an engineer and say why don't you make me an MPEG encoder. You have to have specifications. You have to know that everybody in the room is talking about building the same thing. Or, you might have the prow of a ship and the back end of a bicycle. I found that by putting the work in up front, it was so much more efficient when I would meet with my team because they knew exactly where I was coming from."

Opening Up Creativity

Jeanette is convinced that creativity can prosper in an organized setting. "I am a real strong believer in 'measure twice, cut once.' I've worked with a lot of creative people who feel like, 'well, if you do all of this planning, it takes away from your spontaneity and you can't be as creative.' But, I have found the exact opposite to be true. If you plan meticulously, by the time you get ready to shoot, if everything that you need is working the way that you need to, you have the freedom and luxury to be creative and spontaneous in the moment. There are two kinds of creativity. There is the 'this is really good but how can we make it better' and there is the 'boy, I really screwed this up, how am I going to fix it?' kind of creativity. I think a lot of times beginners get those two kinds of creativity confused. But, I think I would much rather have the former kind."

Jeanette points out that the audio and video capture process is essentially unforgiving. If these key processes aren't planned well ahead of time, the course of the project could unravel very quickly.

Prototyping for Proof of Concept

Because Jeanette was employing a new concept that she had not seen implemented anywhere else on DVD, she felt that it was important to set up a structure and demonstrate that the story feature in DVD Studio Pro operating the way that she expected it to.

"I talked to some experts," Jeanette said, "but they hadn't really done what I was trying to do. Rather than build a whole DVD around this concept and then find out that the concept won't work, I just did some basic tests. A lot of times the marketing literature will say that the software does something and it will be somewhat misleading. Or, maybe you just misunderstood the description. I think that looking at a set of specifications on a piece of equipment or software without trying it and then basing your whole project on it is a form of project suicide."

This step becomes even more important when you are building a DVD for a client. If you are creating a DVD for yourself and you make mistakes, you can usually find ways to circumvent the problems. If you are doing a DVD for a client, and you get in a position that you can't deliver what was promised, your contracts can quickly go out the window. As a developer, you may wind up making a lot of changes without getting reimbursed for your time or effort.

More Preplanning

Pep talks and exercise segments followed a script that Jeanette created well in advance of any production getting started. "The script went through several revisions for the talking head parts," Jeanette said. "I had a set design that I worked out with a set designer. I had a lighting design. I had costume design. In my binder I have the rights to all my music, photocopied three times."

The music was licensed from a music company called BK and Howe, well-known for their specialty, music for aerobics. Jeanette requested a custom mix. "I had to know how many minutes of music I needed," Jeanette said, "what the beats per minute had to be, what the ramp-up and ramp-down had to be. In terms of the workout, I had to know all of that before I could order the music and, of course, I had to order the music before I could design a lot of aspects of the project."

This meticulous approach may sound unnecessarily exacting, but Jeanette understands how tempo is the cornerstone of any aerobics workout. "If you are there with your exercise group and you haven't practiced to the CD, and you find it is ten beats per minute too fast or ten beats per minutes too slow, your whole workout is thrown off. You're either working too hard or not working hard enough. Ten beats per minute in aerobics is a pretty big deal."

The production book serves a secondary purpose, beyond just coordinating the tasks for the project. "This book stands with my DVD as a testament to my ability to put a project together," Jeanette said. "So, when I go to look for funding on the next project, I can show the financiers, here is the output. But, I can also show them the path I followed to get there. They can see that it wasn't a fluke. It should be clear that the project was carefully laid out." Once this working methods are evident, Jeanette believes, the prospects of gaining funding should be much easier to achieve.

Charting the Timeline

Jeanette used Microsoft Visio and Microsoft Project to create the Gannt sheets that helped organize the project tasks. "I basically planned everything I could," Jeanette said, "When it was just planning, it was only me and I was only burning my time. I think that the people who worked on this project with me would work on my next project, because I didn't waste their time. But, I have personally worked on other people's projects where they weren't organized and they didn't know what they were doing. You would show up for a shoot and for the first four hours they were maybe loading the truck and looking for the make-up and doing whatever. The next time they ask you to do a project, you find you are not exactly jumping up and down with anticipation."

Starting from a Technical Background

Jeanette had one big advantage over others launching a similar type of project: experience and familiarity with much of the key technology. "I've been involved in the technology for a long time as a marketing professional and product manager," Jeanette said. "I was familiar with a lot of the intricacies of the technology. I had access to some very cool equipment that was just lying around and got this idea. Initially, I put together a huge budget and started talking to people about it and it was too much of a risk for them—to think about putting $50,000 into this project when I don't have a track record. So, I basically went to Borders and the Library and got *Final Cut Pro for Dummies* and got *DVD Studio Pro for Dummies* and thought, 'Well, why I don't read these books and see if I can figure it out.'"

"I went about it in a very methodical way," Jeanette said, "I got these tutorial books and I did the tutorials from beginning to end with the idea that I would learn the basics before I sat down to do my project. I wanted to know my way around. I spent a week learning Final Cut Pro and I spent a week learning DVD Studio Pro, which in the scope of a lifetime, isn't very much."

Most users probably couldn't learn these programs quite so quickly; Jeanette had the advantage of having worked with multimedia and NLE's before. "I had the general concepts down," she said. "Then, you have to learn the idiosyncrasies. Most Apple programs are pretty easy to learn. I was pleased—they are pretty sensible."

From Technology to Aerobics

The initial project idea came from Jeanette's background and interest in aerobics. "I have been teaching aerobics for many years and I have completed a marathon and a triathlon and a number of different sporting events. One of the things that I noticed as an aerobics teacher who specializes in teaching beginners is that most exercise tapes and DVDs are not really designed for beginners. They are not designed for the person who hasn't exercised in a year or ten years. I learned from teaching my class, where I had a lot of people come into my classes who had not exercised in decades or maybe had never exercised. There are certain things that you have to do on the approach that allow a person to feel successful."

The slow ramp-up in the workout sessions is critical. "If you have got somebody who has never done an aerobic dance class before and you throw 25 dance moves at them in their first class, they are not going to feel like a success," Jeanette said. "And if they don't feel like a success, they are not going to do it again, because it is a miserable thing to feel like a failure. Often, in my classes, I would teach just a couple of moves. I found that the other people in the class were very patient, because they had been beginners once, too. That is the idea. If you look at the American Heart Association or different health organizations, they strongly recommend starting slow."

Consulting the medical profession about the right way to present exercise material was also a key part of the planning. As Jeanette explained it, "I talked to a cardiologist. I talked to orthopedist. I talked to a couple of different medical doctors for advice on how to put this thing together. They gave the thumbs up to my idea. Then I just had to figure out how to make it happen."

Fitness or Weight Loss

The body image aspect of her program is also a integral part of the approach. Jeanette doesn't want intimidating, muscular instructors presenting the workout, but simply normal people.

"I think that is an important point to make," Jeanette said. "In order for fitness to really be successful, it has to be somewhat divorced from cosmetics. A lot of people who take my classes are not going to lose 50 pounds in the first six months. They can exercise until they are blue and they are not going to look like Cindy Crawford. There are other concrete benefits that come of it, like lower blood pressure or better sleep patterns. Or, I have had students in my classes who didn't have to take diabetes drugs any longer. There are very specific things that you gain, but the whole cosmetic thing is so subjective and the media expectations that people have for themselves are so unrealistic that very often people feel like a failure, when in truth they are very successful."

"It is interesting too," Jeanette continued, "because I want exercisers to think like consumers. I want them to think about my product as they would think about any other product, like a muffler for a car or a toothbrush. Does it work? If it doesn't work, it is not necessarily their fault. It may just not work. A lot of exercise gurus develop ineffective programs and then blame the user. It is a very self-defeating approach."

Mixed in with the workout content, Jeanette includes specific challenges on the DVD. One week there may be a challenge to eat more fruits and vegetables. The next week the challenge might be to go out and find yourself a reward for the work accomplished during the week. "There are life challenges that come along with the package," Jeanette said, "that improve the overall health of the person."

Jeanette laughed at the notion that her program bears some resemblance to the progressive approach of Doctor Andrew Weil. "I don't think Andrew Weil dressed up like Carmen Miranda, which is something I did. That is the other thing about this DVD, it is fun. Life is so serious, and if the workout is not fun, you're not going to stick that DVD in there. Not if it's boring. I put a lot of effort into making it silly and making it fun. And making it interesting."

Stretching a Budget in Many Directions

To make a limited budget go farther, Jeanette relied on the age-old technique used by independents, bartering her PR and writing services in exchange for prime requirements of the production, such as the lighting setup by Rick Vaughn. Getting a professional to come in and set up the lighting can do a lot to optimize the quality when you're working with a format, DV, that many consider to be a notch below professional caliber. In truth, well-lighted, well-exposed DV footage shot using classic video production or cinematography techniques compares favorably with many of the accepted professional-caliber format.

"Whether or not you shoot inside or outside, you need somebody working on lighting," Jeanette said. "If you are working outside, you need reflectors. You just need basic things and you need somebody watching the lighting to make sure that your light remains consistent throughout your shoot. So, shooting outdoors is theoretically easier, but impossible to control. If it gets cloudy and you started out sunny, you can't cut back and forth between those segments. That can be very frustrating. I never considered shooting my project outside. I wanted the ability to control the lights and I wanted the ability to shoot when I wanted to shoot, not when the weather cooperated."

Under the Lights

By any estimation, professional lighting generates a lot of heat, but when you have to shut down the air conditioning in the room you're taping so that it doesn't affect the sound, the atmosphere rapidly soars to tropical temperatures.

"I had my makeup done by a guy who specializes in drag queens," Jeanette said. "We actually don't look too sweaty, but we were very sweaty. Yeah, it was very hot. If I did it over again, I would definitely do it at a different time of the year."

All of the video footage was captured in a single shoot using two Canon GL2 cameras. Jeanette explained that one camera was set up high and another was positioned at ground level. "We had somebody doing live switching and ISO's,"Jeanette said. "My friend and co-producer, Donna Donahue, had done a lot of work with video production. She came in with some equipment and I had some equipment and we rented some additional equipment. We, basically, just pulled it all together and did it."

Recording the Audio

Wireless microphones were used to handle the miking of Jeanette's exercise instructions during the workouts. Recording was accomplished with the cameras and microphones being routed into a DV deck through the switcher. No problems with RF interference were encountered.

"You have to be careful when you use wireless mics that your batteries are fresh and that somebody is checking them from time to time—monitoring to sure that the sound is OK. We were with the wireless mics since it was a controlled environment and we shut the air conditioner off during taping. There was very little interference noise."

The video content includes about 50 or 60 minutes of exercise and then what Jeanette calls "the Peep talks," which are short motivational messages. "There are about another 25 minutes of Peep talks organized into weeks," Jeanette said. "Every week you come to a screen and it says, for example, Week 3. You see three buttons: Peep talk, Steps, and Sweat. The Peep talk is the motivational rah-rah speech for the week with your personal challenge that you are supposed to do. Steps is just a video voiceover of the feet for the new step that week. If it is a grapevine, it shows you how to do it. You can do it over and over again until you're sure. When you push the Sweat button, the video displays your whole workout for the week with the warm-up and cool-down and exercise section."

Authoring the DVD

Some of the final touches in the project were accomplished at the authoring stage, where Jeanette applied the story markers in the approach that she had previously tested using dummy video. The video was recorded and edited to a segment approximately sixty minutes long. "I did the Week One segment," Jeanette said, "and I said, 'Week One people, bye bye.' After I said goodbye, I would put a story marker there. Then Week Two would go to the next story marker. Week Three goes to the next story marker. The video sequence is only laid down once. Week One is close to 15 minutes. Week Twelve is about 50 minutes with warm-up and cooldown. It is a nice gentle progression. It is just a few more minutes each week."

Even beginners should have achieved a basic level of fitness after 12 weeks of workout. Jeanette said, "To be able to do a fifty-minute exercise class is good. It is not a marathon. But it is good. It is functional fitness."

Disc Two of the series might venture into new territory, Jeanette thinks, instead of strictly aerobics. "I'm looking at *The Fat Chick Pumps Iron* and *The Fat Chick Chills* out with more yoga stretching and relaxing kinds of focuses. Because those are really the three components of fitness: aerobics, flexibility, and strength."

The Value of Market Research

Focus groups and conventional marketing research may not yield the kinds of information needed by a small-scale DVD producer, as much as just meeting and talking with potential customers. "I don't believe in a lot of the schmantzy things that ad agencies do. But I believe you should find a bunch of people that you think are your target customers and you just talk to them and ask if they're interested in what you're doing. Ask them: what should I call it? What would you pay for it? If you have more anecdotal market research that involves going out and spending time with your customers, you can learn some very valuable things."

"Everyday people can give you ideas that you might not have thought of and they can give you a perspective. You can get very isolated working on a project. While I think market research can be misleading, I think that the other approach, 'if you build it, they will come,' is a recipe for disaster. Whether you are spending a little money or a lot, you are spending money and spending time and you are spending your soul, you might as well spend it on something that is going to sell."

Working as a Product Manager

Jeanette admits that her product manager work is even more meticulous than the work done on your DVD project. Most of her product manager experience has been gained working for Heuris. A good deal of it involved the flagship product of Heuris, MPEG Power Professional. Learning the nature and properties of MPEG-2 video, of course, is a strong foundation for any kind of work with DVD content. Jeanette also believes in the adage that you only have one chance to make a good first impression. In other words, test products and make sure they work as well as possible before releasing them to the public.

"The Heuris MPEG Power Professional product is very sophisticated," Jeanette said. "It has a gazillion features. We took product testing very seriously, so our test harness had hundreds of pages of tests that had to be performed. I certainly didn't perform them all myself, although I managed the testing process. You know, it is so important not to put junk out in the market. You really only have one chance to build a good reputation. And once you get a reputation for putting garbage out there, people don't want to buy your product again. They don't even want to give you the benefit of the doubt."

Maintaining an effective approach to technical support ranks close to the top of Jeanette's considerations for rating the quality of a product. "One of the things that I was always very proud of is that Heuris provides really good technical support. There are people on the phones that you can talk to, and the technical support people are involved in the testing process. This means they are motivated. If people are really mad because something doesn't work, technical support as a group is going to hear about it. So, it is really good to have them involved in the testing process, because they have some incentive to make sure the stuff works."

Jeanette used MPEG Power Professional for the MPEG-2 encoding on her DVD and considers it a very stable product. "When people are evaluating DVD production tools in the purchasing stage," Jeanette said, "it is very important to evaluate the technical support that comes with those tools. There is nothing that is worst than being three quarters of the way through a project and hitting a brick way technologically with a product. If there is nobody to support you or if you are sending your stuff off to the great "404 File Not Found" neverland of technical support and nobody is emailing you back, you might have to have a mid-course correction and buy a different product."

The problem, Jeanette points out, is that often there is no live technical support with some of the very inexpensive software products. There are a number of products available, even freeware products that can theoretically perform tasks central to DVD production. But, sometimes you need an answer on the final day with a deadline looming and products without support can lead a developer to an unyielding brick wall.

"I think a lot of people don't take that into account in the purchasing stage and then they get in trouble," Jeanette said. "Heuris is more expensive then some of the other products that are out there, certainly, but beyond the advanced feature set, they stand behind their products. I've known engineers here to do a special build to get somebody back on the road for a deadline."

So, the key message is that when you are considering product features and prices, take technical support into account, especially ensuring that you can get the support that you need in the final crucial stages of a project. "Your time always gets compressed," Jeanette said. "So, maybe if you are doing graphics development for the DVD, those tasks happen very early in the process, so it is not as critical. But, if you are talking about the MPEG encoding or the DVD burning or the DLT writing, those tasks happen at the very end of the process. Often, people don't save enough time for those elements and that is where they get stuck. Right at the end."

Test as much as you can before you get to that final stage, Jeanette advises. "If you are making a major DVD and you have never written a DLT before, test it well ahead of time. These kinds of tasks are very important components. Before you submit your master to a replicator, ask them if you can do a three-minute test. Put your material on the tape and have the replicator look at it and make sure they can read it and the format is correct. There is some weird stuff that can happen at that stage of DVD production, during replication. Actually, most companies are usually very accommodating. They would rather have you do a test than have you pay for a three-day turn rush job and get in the middle of it and find out that they can't read your graphic files for the face or they can't read your DLT files because something got corrupted. At that point, there is only so much they can do."

Promoting Your Product

Promotion is a demanding, but necessary task, and when you are starting out, nobody is willing to do it with the attention or dedication of the product creator. "You have to try everything you can," Jeanette said. "You don't know which area is going to work. I would certainly hate not to have a self-publishing plan in place. If nobody picks my DVD title up, that doesn't mean I am going to lie down and play dead. I will just distribute it myself and it will be slower process."

"Every time I send it to somebody," Jeanette said, "I learn something that helps me refine the whole project. Even if nobody picks it up, I will have learned things that make my self distribution more profitable. It is a learning process. It is painful to get out there and shill, but nobody else is going to do it for you. There are people who think some magic sales fairy is going to come. Or, they try to get some friend who claims he knows sales to do it. Nobody is going to sell your baby like you are going to sell it. It is painful and sometimes it is depressing when you get turned down a lot, but you just have to get out there and do it."

Managing MPEG-2 Assets

One area of confusion that can be vexing to DVD developers is the extraction and re-editing of MPEG-2 files that are stored on disc. As a part of the authoring and pre-mastering procedure for any DVD, prior to burning, the video content is converted to encoded to MPEG-2 format at a bit rate that is determined by the project settings. Audio and video content are also multiplexed together, the audio converted to Dolby Digital or MPEG-2 audio files, the native storage format for sound on DVD.

Tools exist that can demultiplex the audio and video content from the DVD, but you then have files that are considered elementary streams, unsynchronized audio and video content. Can you re-edit the material at this stage? The answer is not a simple yes or no, but more of a qualified 'maybe.' Non-linear editors, such as Final Cut Pro and Vegas Video, differ in their editing capabilities, partially due to the complexity of MPEG-2 video content, which consists of I-frames, P-frames, and B-frames, each of which relies on surrounding frame content to fully represent the video motion. In her work with Heuris, which includes familiarity with Xtractor, Jeanette is fully aware of the problem and she offers some words of caution to anyone who wants to re-edit MPEG-2 video that has been demultiplexed from a DVD.

"We find most often that people use Xtractor for those times they are doing materials that don't have audio sync. They just take their video, without re-encoding it, and just put new audio to it. Any time you try to take an MPEG file as a source in a NLE system, you are taking a little bit of a risk. It depends on how well the NLE's adoption of MPEG as an import standard is. One of the things that you have to realize is that MPEG is not like Motion-JPEG, where every single frame is there. There are I, P, and B frames. If you do an I-frame only encode, it is relatively easy to edit that material, but you are going to have a very high bit rate, because you don't get as much compression with an I-frame only encode. Most people tend to do a typical I, P, B encode and that means that trying to get your video cuts to fall at the right point in the GOP (the group of pictures) can be a little tricky to implement. I think that is the problem that some of the NLE's are struggling with now. "

After pausing to consider the issue, Jeanette continues, "I guess the real issue is: MPEG was never really designed for editing. It was designed for compression; it was designed as a finishing tool. And, so, I think that some NLE's implement the tricks to turn it into an editing format better than others."

Industrial Edits

This does not mean that editing MPEG-2 files is impossibly difficult, but only that any editing has to be done with an understanding of the relationship of the interrelated frame content in the video file. "The safest thing to do if you are going to edit after using Xtractor is: go in and maybe lop a chunk off the front or the back of the video. That is pretty basic and I have seen that done successfully frequently. But, when you pull MPEG-2 video into an NLE and you actually start putting in new transitions and new special effects, it can get pretty hairy. I'm not saying that it doesn't ever work, but even if you get it to work one time in one set of circumstances, there are so many variables that if you change one little thing, it may not work the next time."

Xtractor, Jeanette tells me, is typically used for synching up a new audio track with MPEG-2 video that doesn't require extensive editing. "Often in an industrial situation," she said, "you decide to do a new voice and the video stays the same. That is what we see people using it for, or maybe they create ambient music mixes and they want to put a new music track in there. But, how well this approach works when importing into an NLE is a function of how well the NLE works."

Video Storage Strategy

If you can't extensively re-edit the MPEG-2 files that appear within a DVD project, what is the best strategy for preserving the video content so that you can re-edit it. "With the DVDs that I have produced myself, when I finish my Final Cut Pro project, I always save my project files. If I have to change something, I don't want to extract the MPEG-2 file. What I really want to do is go back to the original Final Cut Pro project, bring in that captured video content, make my edits, and then re-encode."

If disk space is an issue, Jeanette keeps the project files without necessarily keeping all the digitized assets stored on hard disk. If she does have to return to the original video content, she can hook up her DV deck and retrieve any missing assets without having to keep every second of every asset digitized on disk. "That is a pretty safe way," Jeanette said. "Disk space is cheap these days. I just spent $200 on a 200GB FireWire hard drive and I just keep everything on there. Then, of course, I keep all my original DV tapes, because they are all time-coded and I can always go back and get the assets if I need to."

Improving on Good Results

"I was moderately pleased with the results," Jeanette said. "There are always things that can be better. There are some elements that the distributors I'm talking to may want me to reshoot. It was certainly good enough to get the idea across and possibly get some funding for a second round, if that is what the distributors want to do."

Distributors sometimes want to take a very active role in changing the tone or feel of a project to make it more marketable. Jeanette has no objections to doing a version that bumps the production values a notch higher, but she doesn't want to tamper with the core elements of her approach. "I would be a lot more concerned about somebody wanting to change my script or somebody wanting to change my marketing concept or my title, than wanting to change production values," Jeanette said. "Because that is not my area of greatest expertise. I think, given the circumstances, I really did pretty well. But, I ain't Scorcese."

"I wish that I had used a bigger set, so I could have had more chicklets running around screaming behind me. I wish I could have had more people in the class. I had three other people, but only used two at a time. Three people was about all I could fit on my set. I think it would have been more fun if we had more people and that they were all miked and screaming and hollering and whooping it up."

"And I did learn a lot," Jeanette said. "There are definitely things that I would have done differently. I don't feel as though it is a failure. I feel that for my first time out of the chute, it is pretty good. It is not Oscar-winning material, but for DV and for zero budget, it is pretty good. I think that every time you do a project, no matter how well you plan you are going to learn things for your next shoot."

The project has made Jeanette reflect about her career direction. "I am good at PR," she said, "but it is not my vocation. I feel very passionate about exercise for people of all sizes. And, I feel very passionate about confidence for people of all sizes. And, health for people of all sizes. I think there is a great need in the world for people to pave the way for having real-life health. And that is where I would like to focus a lot more of my energy."

You can read more about the project at *www.thefatchick.com*.

14

Case Study: Producing a DVD Magazine

The merging and recombining of different forms of media have become a natural outgrowth of the DVD phenomena. With the capabilities of a disc that can hold literally any kind of digital content, enterprising producers have begun to push the edge of the envelope and explore new means of communicating. One producer of videos for custom woodworkers, SoftWerks International, latched onto the idea of reducing their production costs by turning to DVD. Their popular woodworking periodical had an established and appreciative audience, but the linear limitations of a videotape cartridge and the expense of producing large, complex projects using analog editing equipment hampered their objectives. Adopting non-linear editing tools to create higher quality productions and converting project plans to Adobe Acrobat for delivery on DVD both helped them reach a new and growing audience and provided a more effective means of conveying the construction details of projects. This approach also opens up the prospect, now being explored by the company, of taking their production to the broadcast medium with a regular television series.

Chris DeHut, one of the founders of the SoftWerks, uses his own woodworking studio to videotape many of the projects. The DVD magazine, *Woodworking at Home,* is generally styled in the manner of the television home improvement genre, but it also bears many individual touches and takes full advantage of the interactive capabilities of DVD.

Overcoming the Jinx of the CD-ROM

In the hey day of CD-ROM development, more than one company tried to succeed in the publishing business by introducing multimedia magazines, embracing entertainment, sports, popular music, youth culture, gaming, and many other energetic topics that could be funneled into a multimedia presentation. Though some of these experiments produced noteworthy results and innovative content, you would be hard pressed to find even one publication that survived that optimistic period. It is hard to imagine that someone contemplating the release of a DVD magazine could look back at that period and still summon the necessary optimism to launch into production. Is it unrealistic to presume that a DVD magazine could prosper when so many CD-ROM magazines met a bitter fate?

When posed this question, Chris reflected, "I am well aware that the CD-ROM magazines that began hitting in the early nineties were interesting, but they suffered a few problems. A lot of them were a port from print publishing into multimedia publishing. Many of them followed that same print format. A paper magazine is a very portable document. You can take it with you and read it anywhere. A lot of CD-ROM publishers took that same concept and brought it into the computer on CD-ROM. You had to sit in front of your computer and read the magazine. Some of them included interactivity and short little video clips."

The group that now composes SoftWerks had thought along those lines, as well, but turned away from the limitations of poor quality video, feeling that the technology was not ready to do top-notch presentations.

"We decided not to go down the CD-ROM road," Chris said. "We researched it very heavily and didn't feel that it offered a viable product to present on CD-ROM as a periodical magazine. At that time, we were actually going to publish it for the precision metal-working industry. Everybody would have had a computer with a CD-ROM drive. The audience was solid, but what we really wanted to communicate, you just couldn't do adequately with the CD-ROM."

The arrival of DVD solved many of the problems that SoftWerks saw with the CD-ROM. As Chris explained, "SoftWerks International had been producing woodworking videos for about two years on VHS videotape. These tapes are very specific. We would design a project and then film the construction of that project and create blueprints. Then we would package the videotape with the blueprints. But, these videotapes addressed single-topic issues. It wasn't a subscription, but the videotapes were sold in retail stores. We wanted to communicate with a much broader presentation on a wider variety of woodworking topics. The only way we could do that was with a magazine-type format where you have a number of different types of articles in each issue and it is on a periodical basis. Every other month, we ship out a magazine."

As SoftWerks produced a string of VHS videotape projects, they began to realize that this medium wasn't the answer. The tapes were too expensive to produce and too expensive to ship. They also lacked another key element that SoftWerks felt was important to their approach: interactivity. Ideally, a woodworker should be able to jump from article to article, to quickly view those topics and projects of greatest interest. "DVD really solves this problem," Chris said, "by letting you present a video magazine in a very user-friendly way. DVD players, both on televisions and in personal computers, have one primary function that is perfect for magazines—a table of contents. Or, what you would technically call a *menu system*."

Viewers of the *Woodworking at Home* magazine can pop the disc into their DVD or computer and navigate to those articles that catch their interest. The expectation for this type of publication is that the audience will actually want to construct the project being shown. Understanding that most woodworkers are not going to have a DVD player or a computer in their workshop, each *Woodworking at Home* disc contains a transcribed version of the project audio and key photographs that illustrate the portion of the project being discussed. This information, presented in an Adobe Acrobat file, can be printed and taken to the work area. The more intricate setups and key photographs, as well as the detailed CAD drawings, can be used as a reference as the project construction takes shape.

The CD-ROM had significant space limitations when it came to presenting video, but DVD surpasses that restriction. "DVD," Chris said, "with its 4.7GB capacity, allowed us to present up to two hours of video. We could also present this video in a way that lets viewers easily navigate from article to article. We also could include the paper documentation that was essential to successfully constructing each project. Realistically, the DVD, with its large capacity, extremely high quality video, its menuing system, its ability to store ROM data as well as conventional video data, all came together to fit our idea of what a modern magazine should be."

Handling the Video Production

In true independent fashion, SoftWerks took the approach that they would handle all of the video production inhouse, including the encoding of the video material for delivery on DVD. They took care of all the individual details—package artwork, menu design, authoring of the content, production of a master DVD—to the point where the completed magazine was handed off to a replicator for disc manufacturing. During authoring and production, SoftWerks relied on more than twenty individual software applications, including Adobe Premiere, Pinnacle Systems digital video capture cards, a variety of graphics applications, and so on.

The *Woodworking at Home* approach uses a presentation format very similar to the type that has been refined by a number of home-improvement shows. "*New Yankee Workshop* is a prime example," Chris explained. "But our style is a little bit different: we are a bit more detailed. We don't have the 22-minute time limit imposed on us. If we want to go 36 minutes on a project, we go 36 minutes on a project. I can present much more information. We use staff members, myself included, in the presentation out in the shop. We also have a couple of hired authors and contributors and we hope to expand on that throughout the course of the next year."

Selling the Concept

One of the hurdles faced by the *Woodworking at Home* staff is the difficulty in convincing potential advertisers that this concept is valid and can succeed in the marketplace. "Because print magazines have been around for centuries," Chris said, "getting a couple of advertisers involved is merely a matter of phone calls. You don't have to convince them of the concept. As far as I know, we are the only company in the world right now producing a periodical magazine on DVD. You've got to have a real product. Everyone remembers the dot.com bust. You can sell all the sizzle you want, but it is the steak that people want and if you don't give them the steak, you're going to fail."

Faced with these barriers, it has been an expensive proposition for SoftWerks to land new customers. The business plan is counting on getting through the first year successfully and building momentum on repeat customers. Advertising and promotion are expensive but necessary parts of this process and Chris feels strongly that this is an approach that can flourish as more customers become familiar with what they have to offer. In one of those unpredictable quirks that characterize any product marketing effort, the *Woodworking at Home* DVD has proven extremely popular in the Japanese marketplace, where there is a strong interest in American woodworking techniques. The momentum continues to grow domestically, as well, but it's still early to determine whether the concept will prosper long-term.

As the subscriber base has increased and the projects have gotten more polished and refined, Chris has found it easier to line up sponsors and this has brought a valuable stream of both revenue and equipment to the fledgling production. Delta International Machinery recently signed on as a sponsor, providing all of the stationery machine tools and all of the bench top power tools. Viewers see these tools being used in each of the episodes and this exposure can help Delta International Machinery gain new customers.

"In sales," Chris said, "demonstration is very important. Especially with high-end woodworking tools, you have to demonstrate your product to a customer. In every issue, I am demonstrating a wide variety of the Delta tools and products. So, they get a constant stream of demonstrations. That is, literally, our advertising mechanism. We can demonstrate your product and you will get third-party credibility to your sales statement and you can show potential customers how effective the tool or product actually is."

Sponsors, of course, do whatever they can to try to maximize their exposure. "They also wanted to hang a large banner in the shop. I compromised with a smaller magnetic sign that would hang on the shop door behind one of the machines."

Customer Response to Advertising

While magazines and television and other forms of entertainment often require advertising to support their productions, many audiences have a clear aversion to excess advertising. Chris was concerned about this factor and determined to handle sponsorships in a very unobtrusive way. "We have actually gotten compliments from some of our customers. This was rather interesting, because I was very hesitant to put advertising in the magazine, but I knew that in order to survive, we would have to have advertisers. Right now, the way we are doing it, the sponsorship messages are about 20 to 30 seconds long. It is just a still image with a quick voiceover. We have about five or six sponsors right now, so it is very minimal, but what we trying to do seems palatable to the customers. First, we don't advertise and we don't pursue any advertisers for non-woodworking-related products. In other words, the advertising is part of the editorial content. We are not advertising vitamins in a woodworking magazine. Second, I put advertisements in where they fit the tasks being shown. For example, there is one device called the Gripper. It is a push block that they use on one of the machines that acts as a safety device. In this two-hour magazine, I use this product throughout, but wherever I use it the most, where it gets the most camera time, I put their advertisement in at the end of that story segment. Our customers say, 'we like the way you advertise because what you are advertising is relevant to the editorial content.' In the case of the Gripper, a number of people had said, 'I saw that thing and I was wondering what is was.' And, of course, the advertisement told them more about it."

Subscribers asking questions before signing up for *Woodworking at Home* often ask how much of the magazine is advertising. Chris has a ready answer. "Right now, we humbly say, it is about a minute and a half. Our audiences don't want to watch a half hour of commercials. That is about what you would see watching two hours of television. When you minimize that and when people see how minimal the impact of advertising is on the production, they are grateful for it."

Chris admits to struggling for the first few months of production to establish a working business model for the magazine. "Frankly," he said, "for anyone considering a magazine production, this has been a very tough learning curve. For anybody who is in print publishing, advertising is easy. For video, it is a whole different ballgame. Sponsorship and advertising is a very important aspect for anyone considering this type of publication."

Knowing that digital video production tools offer the opportunity to produce video spots for a fraction of what conventional video production costs makes it possible for Chris to envision more sponsors jumping on board as time goes on. But at this stage, as it is a pioneering effort, sponsors typically want to see results. So, the staff at *Woodworking at Home* is actively working to demonstrate the kind of acceptance and results to bring sponsors on board.

Using Acrobat to Present Plans

One of the key elements of the *Woodworking at Home* approach is the inclusion of detailed project plans, delivered in Acrobat format on the DVD itself. This trend follows the approach that has become common throughout the computer industry: providing product documentation in electronic format rather than as a printed manual. As Chris says, "From a woodworker point of view, I'd rather see that paper pulp go into a piece of furniture instead of going into something that is going to end up in the dump."

The universality of Adobe Acrobat solves a number of production issues. "For us," Chris explained, "the Portable Document Format (PDF) lets us present a variety of graphic elements. We have, of course, text—formatted text. We have photographs, The key element is our CAD drawings. We couldn't publish our CAD drawings in native format and expect everybody to go out and buy a $3000 CAD program to view them. The PDF format allowed us to put all of these elements together into one piece and then publish in a universal format."

The intricacies of presenting sufficient graphic details to ensure the provided plans would be usable actually proved easier in electronic format than in a typical printed approach. "In a CAD drawing," Chris said, "you might have thirty or forty components displayed. In a print magazine, you would take those components and magically dimension them on one or two 8.5x11-inch sheets of paper. The results are horrendous to read and interpret. Even in our old project plans, which we sell with our videotapes, we have a printed blueprint that is two feet by three feet. We still have to cram all that information on one sheet of paper. With the PDF and CAD system, we can literally put one component or two components per sheet of paper and add tremendous detail with large letters that everyone can read. In reality, publishing the CAD files in PDF allowed us to put more detail into our drawings and give our customers more clear information than we could provide in any other media."

Spinning Off Segments for Broadcast Use

Digital content is malleable, making it far easier to repurpose than equivalent analog material. Producers of digital video content can immediately begin thinking about other avenues for delivering their content, such as streaming segments from the Web or submitting content for television broadcast. This process is simplified by the fact that MPEG-2, the compressed storage format for DVD video, was conceived as a television broadcast format, making it a viable medium for submitting material to broadcast stations.

Chris admits that SoftWerks has considered these options. "We were looking at all the possibilities," he said. "When we were in production of the videotapes, realizing that we wanted to expand on the content, we approached several networks about a television series. Unfortunately, given the current economy, networks such as DIY and HTTV are running more advertising on their shows advertising their other shows than they are trying to take on more expense and more product. It was just a matter of timing. We believe in the years to come, once we expand our audience base and expand on this concept of the video magazine, we will be able to repurpose much of our video and put it on television."

Following One's Passion

As anyone who works for a living knows, the source of one's livelihood doesn't always correspond with interests or passions. Chris found this out over time while doing software development and consulting in the precision metalworking business. "That was my vocation," Chris said, "but my passion is woodworking. Three years ago, my partners and I decided to start backing away from the manufacturing industry and begin developing products with a company that feeds the soul—to work with passion. We started producing the woodworking videotapes and that led us into the magazine that we're now producing."

The same passion for woodworking is shared by anywhere from two million to seven million people in the United States, according to whose estimates you believe. Chris is more conservative in his demographic estimate. "I believe the true audience is somewhere around a million or less. For paper magazine subscriptions, the top-selling magazine is around 550,000 subscribers annually. From the top dog, they go down to around 150,000 at the bottom end of the market. Our ultimate goal is to gain 500,000 subscriptions, but we are fighting a couple of factors. We will have to wait. Depending on what numbers you look at, 43 to 55 percent of U.S. households already have a DVD player. We've got access to a half market. Everybody can read a magazine, but the only way you can view our video magazine is to have a DVD player. We have to overcome that, which right now is a function of time."

Chris notes that major videotape outlets, such as Blockbuster, are phasing out videotapes and bringing in DVDs. Part of the challenge, he confesses, will be selling the concept to prospective viewers who aren't familiar with the idea of a video magazine. "We know what the scope of the market is because we've seen what other magazines are capable of reaching: about half a million subscriptions. We are also presenting a video format, so we hope we can grab some of the television audience. We've got to fight the technology until it catches up. We've got to sell like crazy to get people to understand the concept of a video magazine. Our ultimate goal of 500,000 is, I truly believe, an attainable goal, but we know it is going to be a long-term goal."

Adding Subtitles

Chris has done the DVD authoring for the first few issues of the video magazine using DVDit! from Sonic, but he has recently run into an obstacle that is leading him towards investigating other authoring solutions. "What is happening here," he explained, "the Japanese market is getting really strong. Many of our Japanese customers are requesting that we put subtitling in our videos. Unfortunately, DVDit! doesn't support subtitling. Here in the U.S., we have hearing-impaired people who would like subtitling as well. It is a feature that I would like to add to future editions. Because we go through the process of transcribing nearly every word in every video, we already have the text to produce the subtitles."

Although subtitling is a feature that appears in Apple DVD Studio Pro, Chris isn't ready to switch platforms to gain this feature. Authoring applications on the Microsoft Windows platform that include this feature are generally limited to the higher end, more professional products.

Building a 17th Century Hall Table

As an example of the content provided in each issue of *Woodworking at Home*, a full project video is included on the DVD with this book detailing the construction of a 17th century hall table. Even non-woodworkers may find it interesting to follow the crafting of raw wood into a piece of quality finished furniture. This sequence also provides a glimpse into the workshop that Chris uses as his studio and illustrates some of the problems that he faces in videotaping and presenting these projects.

Chris carefully considers each of the projects based on prospective audience, furniture type, and degree of difficulty. "Right now," he said, "when I'm picking content for the magazines, I am not completely sure about the nature of my audience. I think I know who is buying the magazine, but I don't really know for sure who is buying. Among the different age groups, younger people tend to favor more contemporary art-type furniture. Your older crowd tends to favor the furniture they are familiar with, such as the conventional early American Shaker or mission-style projects."

During this period while the video magazine seeks its voice and approach, Chris leans towards an eclectic choice of projects spanning every conceivable type of project. As he explains it, "Right now, we are trying to put everything in the magazine—every period of furniture, every style imaginable, different rooms where the furniture would go. Big projects. Little projects. A very, very diverse story selection and project selection."

The hall table represents the oldest designed piece that Chris has ever constructed. From this wide selection of projects, he hopes to get feedback from his customers and get a more accurate read of their preferences. With the continually expanding customer base, this should provide enough useful feedback from the first year of operation to shape the direction of *Woodworking at Home* in year 2.

A Project Idea is Born

"On this particular project," Chris said, "I was researching antique furniture at the library, looking at different periods of furniture to pull out design elements. I wasn't necessarily intending to go and replicate an antique piece—I just happened to stumble across that hall table. It had the right shape, proportion, and scale. It looked appealing and it looked like it could fit into any home. The other side of the coin was: I happened to have some very highly figured wood and just enough of it to make a project of that size. The two pieces really came together. The table was something that I wanted to make out of that wood."

Figuring out how to construct a piece, particularly if you don't have anything to work from but old photographs can be a challenge. Chris explained that he often works from something as simple as a photograph. First, he would find a style that he wanted to do. Then he would study whatever source material was available to devise the best way to do it. "Usually in old furniture books," he said, "you will have a black-and-white photograph, usually horribly grainy with very poor contrast. If you are lucky, the book will give the length, width, and height dimensions. So, from that photograph and from those dimensions, you figure out how to do the rest. We take those basic elements and then start designing the project in the CAD software."

Joinery Methods

Surprisingly, the joinery methods that were used more than two centuries ago weren't that different from the ones used today. "The tools that we use to make those joints are modern," Chris said. "We are doing the same thing, except with electric tools. I personally don't believe that you have to recreate a piece using the tools that were used back in the seventeenth century. That doesn't add value to the piece."

Chris also believes that his audience benefits from discovering a variety of different approaches to accomplishing the same task. "Part of our editorial mission," he explained, "is to show our audience the variety of ways to accomplish a task. We have found this is often missing in print magazines and television shows. On one of the popular television shows, *New Yankee Workshop*, Norm always does his mortise and tenon joints the same way and he has been doing them the same way for 15 years or so. There are dozens of other ways you can do those same type of joints. What we try to do is different. In one issue, we will do that joint using the most modern tool. The next issue, we will show the people how to make a jig and use another tools to mill out the mortise. Then, in a later issue, when we come around to that joint again, we'll demonstrate yet another technique on how to do it. Over the course of time, we can show our audience a wide variety of techniques to accomplish the same task."

Through this approach, woodworkers viewing the magazine gain a breadth of knowledge that serves them well on future projects. This approach does assume the audience is equipped with most of the primary woodworking tools. "We expect they will have a table saw. A jointer. Some of the basic primary tools. Those can range in price; the table saw that we use is a $1600 saw, but you can buy one at Home Depot for about $150 that you can get by with."

Creating CAD Project Plans

Using a CAD application, Chris takes the general shape of the furniture piece and draws that in as an outline drawing. Then, using standard woodworking joinery techniques, he starts applying how each one of the components will fit to the next component. Once all of the details have been added to the drawing, he revisits the dimensioning of the piece. When complete, the document gets published as an Adobe Acrobat PDF file.

"In this last issue," Chris said, "we had included a larger project—a wardrobe style armoire. Just in part one of the project, the text and all of the drawings filled 41 pages. That is about the typical length of an entire woodworking magazine. Using Acrobat, we can provide so much more detail than is practical in a conventional print publication."

Working without a Script

During the CAD work, Chris considers two aspects of the project. "One is," he said, "how am I going to make it? Two: how can I make it and feature a technique that we have yet to talk about. Reflecting on what I said earlier, we actively try to show different ways of doing the same thing. During the design phase and the engineering phase of the project, I am determining how I am going to make it and how I can make it so I can convey it to the audience in its simplest form. Then, I want to show them a variety of techniques or put a primary focus on one specific technique—some type of joinery or profiling or something of that nature."

In the methods that Chris and his partners have developed over the years, a script tends to make the video leaden and unconvincing. "To me," Chris said, "a script is probably the worst thing in the world if you already know your subject matter. David Riley and Dick Sing (the other two contributors to the magazine) and I are all teachers. We have all done one-on-one training and teaching and mentoring, showing people how to do things. So we are confident in our abilities and confident in our message. We know what we need to show. We know how to explain and demonstrate a task. None of us are actors."

"Because we are not actors," Chris said, "we can talk confidently and knowledgeably about our subject. We can convey that information comfortably in the terminology—we will call it the technospeak of the audience. It is a very comfortable conversation. Early on, when I was making the videotapes, I tried to work from a script. On camera, it literally looked like I was reading a script. Eventually I abandoned that approach. The new approach is: let Chris be Chris. Let Dick be Dick and let David be David."

Figure 14 - 1 **In the workshop facing the saw**

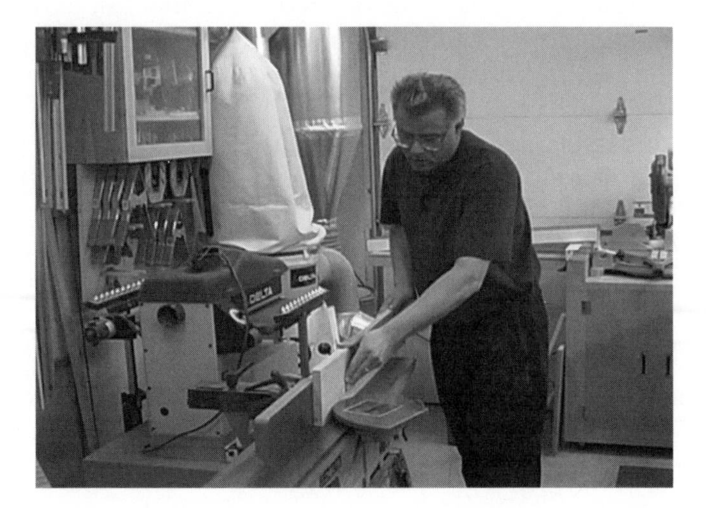

"This is all shown," Chris continues, "literally from raw material all the way through to the host standing next to the project saying very proudly, "I'm Chris DeHut from *Woodworking at Home*. Thanks for watching."

In a conventional production environment—whether television production or commercial productions—a storyboard is almost invariably used to create that storyline. In the case of *Woodworking at Home,* the project that staff members are constructing or demonstrating becomes the guiding factor that controls the storyline. A certain sequence of steps and a particular process must be followed to get from raw material to finished product. That process controls the exact order of everything that appears in the finished video.

"It takes that whole concept of communication a step further," Chris said. "As I mentioned, we started from print, and now we are doing it in video so that we can communicate more effectively. Instead of using an actor, we use real people who communicate properly with the actual subject matter. It brings in another powerful element of communication that does not exist in most other media."

Handling the Camerawork

Chris has set up fixtures and designed an approach so that he can videotape segments by himself. In effect, he serves as the videographer, sound person, impromptu scriptwriter, narrator, editor, stage designer, and wardrobe specialist—all at the same time.

"I do all of the camerawork—100 percent of the camerawork right now. So, when I'm filming myself, I've designed and built a very elaborate camera mount. We've got a lot of obstacles in the woodshop. You've got lights hanging from the ceiling. You've got power cords dropping down. Air filtration systems. Tracks for doors. And, of course, machinery and equipment. A tripod won't work out there. A general gib crane won't work out there very well either. We had to literally build a camera mount that would allow us to position the camera anywhere in the shop to look in any direction. And, of course, on that mount is a television so I can watch myself doing the presentation."

Chris continues, "When I am videotaping the other hosts that do this, they can also see themselves on the TV monitor. They can get comfortable with how they want to look and feel and position themselves. They can also direct me for positioning camera angles. It is a very intense process. Luckily, I've been doing it for three and a half years now—it is second nature. I just know I have to go grab the camera, move it to this point, point it in that direction, snap it in, focus it, bring it out to frame, run back around and I am taping."

For camerawork, Chris uses the standard bearer of the miniDV revolution: a Canon XL1S. "For me," Chris explained, "the only drawback is the door where the tape gets inserted is not gasket sealed. We're in a woodshop where we create tons of dust. Every once in awhile, the dust gets in across the read head and you'll see it as a sparkle or a dropped frame or something. We do our best to keep those out of the production, but it is an ongoing battle."

One-Man Operation

In a single-person production, there is always the risk of losing concentration around power tools, a factor which is never far from Chris's mind. "Any one of those machines out in the shop can remove a finger, a hand, literally in a split second," Chris said. "To further complicate things, on the primary tool—the table saw—because of the audience view where the camera is positioned, you can't leave the guide on. It hides the operation that we're showing. So, we run the saw without the blade guide and this gets my hands real close to the saw."

Figure 14 - 2 **Close up of the hands at work**

The planning, in this case, precedes the execution; Chris has worked out a complete drill to manage the process. "I preset everything so that I know exactly what I've got to do," Chris said. "First, I set up the camera where I want it. Then, I come back to where I need to film from—where my stand position is. I start the recorder. Think through the process a little bit. Explain the setup, the procedure, and then I actually do the cut. During that 30 seconds, I have to make sure the battery in the microphone isn't dead. I have to make sure that the sound level is appropriate. I've got to make sure that the shot is framed correctly. And, I have to make sure that it is in focus."

"I also have to look kind of pleasant while I am doing it," Chris laughed. "I don't want to have a terrified look on my face. Then, I make the actual cut for the machining operation and try not take off a thumb or a finger in the process. As you might expect, it is incredibly intense."

Figure 14 - 3 **More close-up work**

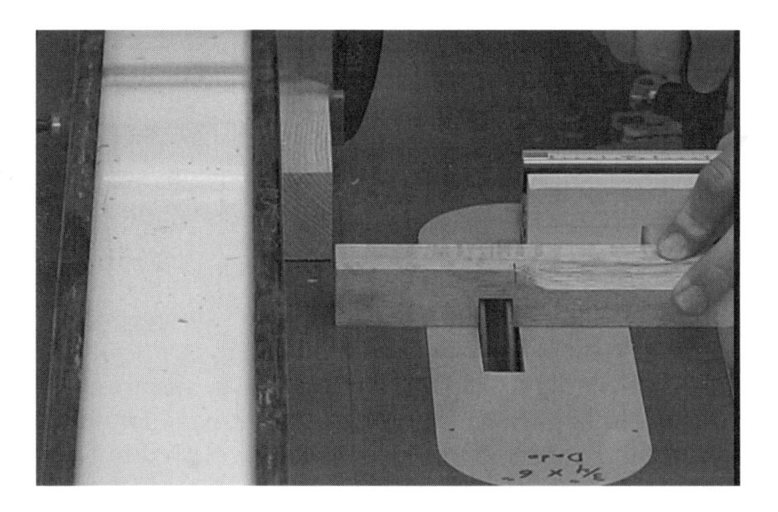

Chris admits that the process used to scare him more than it does now. "The very first few stories that I shot with the VHS videotape, it was terrifying. Now it has gotten to the point that, through experience, many of the tasks have become second nature. Setting the camera positions doesn't even require any thought. Checking the sound level—because you have to hear yourself—you develop a series of pops and whistles. You hear it off the monitor so you know your sound level is right. And, through that process you also know the battery is live."

Wireless Audio

Without a sound technician or boom handler, Chris relies on wireless microphones for freedom of movement and professional grade audio. He alternates between three wireless mics, a Sony, an Azden, and a Radio Shack version, which tend to have very short lifespans because of the difficult environment and dust-filled atmosphere.

"We have pretty close to state-of-the-art dust collection in the workshop," Chris said. "But, you are creating dust faster than two vacuums can clean it up."

The wireless microphones deliver very good quality, even though Chris relies on the less-expensive models. "The model I'm using right now is a pretty cost-effective one. I think it is under $200—very inexpensive. I keep buying the cheap ones because they get damaged out there. The warranty doesn't cover something that got dropped into a machine and chewed up."

Ideally, when things are operating properly, Chris manages to forget about the mic as much as possible. "The mic that I use has a transmitter belt pack. I run the wire up through my polo shirt and put the mic on the V-neck area. Ideally, I try to buy shirts that have the same physical dimensions, because if you get it positioned right, you don't get the swishing sounds. Being up high on the chest, the mic is away from everything so that you don't bump it too frequently. Just through natural positioning, it seems to work out very well."

Fighting Noise

Woodworking power tools are noisy, so talking while cuts are taking place is pretty much impossible. A comment from one of his customers, though, got Chris thinking so he changed his processes to both pick up the pace of the video and contend with the noisy machine environment. "One of our viewers told me, 'I love the magazine, but it is a little bit slow-paced. If you could talk while running some of the machines, instead of talking and then running the machine, it would be a little more enjoyable.' When you are videotaping in the shop, you tend to think analog. First we'll explain it. Then we'll do it. And you just can't talk over some of the machines, because they are so loud. So, I've changed my approach."

"What we do now is this," Chris continued. "I make my explanation up front. Then, once the explanation is complete, we tape the cutting. While editing, I can take the portion of explanation where we are talking about the setup (and I have to show elements on the machine that we are talking about) and play that back just as it was filmed. Then, there are portions where I am talking about what you should do during the cut. We splice in the video footage of the actual cut, turn the gain way down on the machine's noise level to about 5 percent, and overlay the voice segment with the explanation. Once we started doing that, I had to send an email back to that fellow, saying 'Thank you. Great suggestion!' This new approach is working very well."

"Noise has been a challenge for us because of where we shoot," Chris said. "There is a lot of noise outside the shop. A couple of tricks help there: because we are in a shop, we have shop door—one of those big overhead garage doors. We replaced those doors with insulated doors. That kept out some of the sound. Other sounds, though, like when the landscapers show up and run their blowers, can be a problem. We try to run a device called an air cleaner, which is contained in a ceiling mount fan. For our purposes, it cleans the air of the very fine dusts. This creates a little bit of white noise in the background that seems to make the other ambient sounds go away, the leaf blower, the dog barking. You can hide a lot of those sounds with ambient noise. If the blower unit is positioned in the right place, it doesn't get annoying as background noise—it sounds natural rather than annoying."

Changing microphone patterns can also help contend with unwanted noise. "With the wireless microphones we use," Chris said, "we can put our own microphone into the transmitter and each of the different microphones has a different pickup range and pattern and so forth. Even just changing the microphone can get rid of a lot of the undesired noises. That is very important: finding a microphone with the right pickup pattern so you get what you want, rather than everything else in the background."

Shoot for Editing

"One of the books I had read on video production," Chris said, "especially in this modern age where we edit digitally, said: shoot for editing. Don't shoot for content. That didn't make sense until I tied it together with that customer's comment. Let's videotape everything and then, because you can do a wonderful, smooth job in editing, you can overlay and splice and move things around as you need to. Don't spend three hours trying to tape a ten-second segment when you can fix that ten-second segment in the editing process in a matter of minutes."

To produce the 20-minute segment for the table, Chris spent about five days of work from rough material to applying the finish on the product. For each minute of finished video, he shoots about three or four minutes of videotape. "I will probably have an hour or an hour and a half of rough footage during that whole week of work," Chris said. "We edit that down to about a third of that length and it comes out to a 21-minute story."

Figure 14 - 4 **Applying the finish**

Editing and Archiving the Content

Chris does all of the video editing, working in Adobe Premiere. He actually prefers one of the earlier version of Premiere to the later upgrades. "Since we purchased it two years ago," he said, "there have been interim upgrades and in most cases they have made it less useful. I tend to stay with the original version. Doing this editing on a 1GHz machine we're really pushing the envelope."

Archiving also represents an important task in the overall process. "It used to be that I would save all of the rough footage, and then, of course, the final output. We archive two forms of each project. We take the MPEG-2 file that is used on the DVD and store that either on a DVD or a CD-ROM as a native file. Then we also take the DVI file, which is what Premiere uses, and we write that back to the miniDV tape as a completed story. So, we have both an editable copy and we have the archive copy, if we ever had to remake a master DVD."

DVDs as the Future of Publishing

A good deal of risk and pure speculation accompany any project such as the *Woodworking at Home* video magazine. Nonetheless, Chris encourages other developers who are contemplating similar projects to be bullish about the future and to go for it. He firmly believes that DVD and its successors represent the future of publishing.

"I've put everything on the line for this concept," Chris said, "and it is costing us a fortune to get it on the market. I believe you are going to see more and more people publishing other elements of media. Traditional print magazines will start to evolve into the video magazine format. There will be people who instead of publishing a print book will publish an independent film. DVD is truly the future of where everything is going to go. You find DVD on PCs. It is an integrated type of media that can include interactivity. You can put any kind of digital data on disc or run applications from it or use it to interact with the Internet. It is a product that can be viewed easily on a laptop computer or on your personal computer or your television. As a technology, I think DVD is going to bring together many other technologies. I tend to think that DVD is going to be more at the core of business in the future, rather than the way people perceive the Internet as being at the core of business."

Other factors that will fuel the shift to more reliance on DVDs as a communication medium, Chris feels, are the visual orientation of the baby boomer generation and the simplicity with which the latest digital video tools speed up production. "With the baby boomer generation growing up on television," Chris said, "visual communication has been the great communicator for many of us. Even though great literature is still a very important part of our lives, you still find a majority of people sitting in front of the tube. From our perspective at SoftWerks, it is a matter of using video to provide a little more education."

The huge and complex operations that characterize most video and film productions will give way to more flexible, less expensive approaches that can be accessed by any small business, Chris believes. "Looking to the future, it comes down to a very small company that can produce a very high quality product with few resources," Chris said. "If you look at most television shows, the credits list goes on and on. Eighty percent of those people are doing twenty percent of the work. If you look at a small, lean company, such as an independent digital video firm, you've got twenty percent of the people doing all the work. Figure it out and just do it—plain and simple. Does it really require three-hundred people to produce a two-hour video product?"

With the latest generation of digital video camcorders and non-linear editing tools, some pretty impressive work can be done by one person working alone, as Chris demonstrates in the video project included on disc.

For more details about current issues of *Woodworking at Home* and upcoming projects, visit: *www.woodworkingathome.com*.

15

Case Study: Creating a High-Caliber DV Short Film

Proponents of DV technology often argue that the medium opens up moviemaking to anyone who can afford a camera and a computer. Detractors of DV technology similarly argue that the medium opens up moviemaking to anyone who can afford a camera and a computer. Truth be told, miniDV format cameras and inexperienced moviemakers have produced a monumental volume of very bad movies. But, what happens when the time-honored production techniques of filmmaking are applied to the medium? The nature of the story, of course, and the way that story is told represent a key part of the equation. The other side of that equation involves the techniques used to capture light and sound, to acquire images and then manipulate them, to weave music and sound effects into the mix. What if skilled lighting technicians are employed to achieve the optimal lighting in each scene? What if high quality lenses and a cinematographer oversee the production to create the desired visual effects? What if the audio is properly miked, monitored and mixed by audio specialists? Will the resulting DV content rise to professional standards?

Victor Muh, a Paris-based moviemaker and digital video production specialist, thinks that DV is undervalued as a professional tool and he set out create a short movie that would push the boundaries of the medium. *The Chinese Shoes* was videotaped on a Canon XL1 camcorder, taking advantage of the interchangeable lens feature of this camera to achieve exactly the visual properties that he wanted. A Mini 35 Digital lens adapter from P + S Technik allowed high-quality 35mm Cooke lenses to be used in the production. Through collaboration with cinematographer Pascale Marin, Victor was able to manipulate depth of field and focusing more creatively than is possible with the stock Canon XL1 zoom lens. Throughout each stage of the production process, Victor overcame a shifting array of obstacles to achieve results that clearly indicate the potential of DV as a moviemaking tool.

Filmmaking Roots

Victor grew up in Hawaii and California and he gained an interest in filmmaking when very young, prowling through the stacks in local libraries during the summer months when grade school was not in session. He was particularly intrigued by special effects and the techniques used by well-known filmmakers and read everything he could find on the subject. "*Star Wars* had just come out," Victor said, "and I was already a fan of *The Six Million Dollar Man* and *The Bionic Woman*. I read all the books about how movies were made and how special effects were done, but I didn't take it any further through my high school years. When I graduated from high school, my Mom bought me an 8mm Canon video camera. I started to make my own videos that my friends loved, but I really didn't know where to go to develop this interest more."

While still in college, Victor took a trip with a friend to Austria to visit someone they had first met in L.A. "I pretty much ended up staying there," Victor said. "It was my first experience in Europe and I was completely enchanted with Austria. I was reading a lot of the beatnik philosophy and poetry and staying in Austria seemed like the thing to do."

Victor went to the University of Vienna to learn German and then decided to pursue moviemaking by enrolling at the Vienna Film Academy. "As part of the entrance requirements, we had to make a short film based on a generic script they provided. We had to interpret the script in our own way. At this point, I had saved up and bought a Hi-8 Canon video camera and I was going to shoot my entry film with that. But the camera got stolen in Paris. My whole idea to enter the academy kind of went up in smoke. The camera actually got stolen not far from where I live today."

The camera disappeared at the conclusion of a trip to the U.S. "I had all these packages," Victor said, "and I was tired and jet-lagged. I put everything down for a moment and just looked around. When I picked up my bags again, I noticed that the smallest bag, which was my camera bag, was missing. At this point, I had business contacts in the States and I decided, 'If I have to save up again to buy a camera to make my college entrance movie, I might as well try to do some business instead of working odd jobs or under-the-table jobs in Vienna." As a result, Victor spent the next two years in Austria, distributing snowboards and streetwear.

Moonlighting in Paris

Since Victor settled in Paris, he has focused his energies on creative moviemaking, but he also makes a living doing audio-visual work. He produces reports for both American and European cable TV, including CNBC and Eurosport, and making action sports and music videos.

"I am pretty new to Paris," Victor said. "My background is in surf films and surf videos. I've always worked with really low-budget projects. I know how to shoot and how to edit and how to produce and everything like that. All wrapped into one. I moved to Paris do more of my own creative work."

Honoring the Memory of a Friend

"I'm always open to new ideas," Victor said. "This led me to making connections in the surf and the snowboard business in Europe. A friend of mine in Hawaii—Mark Foo—was a very well-known surfer with a TV show called H3O TV. It was a surf TV show from Hawaii that was aired locally and on the Fox Sports Network on the mainland. Mark died while surfing. During his life, Mark always wanted to spread the joy of surfing to the world. I wanted to honor his wish and this inspired me to contact H3O and offer to distribute their show and market it in Europe. H3O agreed and I actually got the first surf show on European TV."

As a forerunner of many TV surf shows that are now popular on European television, the H3O shows proved slightly ahead of their time and only gained a marginal following. Since then, market tastes have shifted considerably. "Actually," Victor said, "the whole surf, skate, and snowboard action scene in Europe has really boomed in the last couple of years."

Victor quickly went from distributing the H3O shows to producing segments for the show. "From that," he said, "I started to make surf videos and to work on a 35mm action sports film that was very similar to a Warren Miller ski film, but it was targeted for the European market. It was called *Nuit De La Glisse* (*Night of Surf*) because they would only play the film at night. It was all action sports: surfing, skating, snowboarding."

Gaining Experience in DV Techniques

The early video work that Victor accomplished was done in Beta SP, but he watched the development of DV technology with keen interest. When the original Canon XL1 camcorders became available, Victor was one of the first in line to buy one. "I have always enjoyed Canon cameras," Victor said. "I don't know if it is their lenses or what it is exactly. The colors seem so much more rich and vibrant than what I see with Sony cameras and the others."

Victor hears people criticizing the color handling capabilities of most DV equipment and doesn't agree with the harsh judgements that are sometimes voiced about the medium. "I personally don't have any trouble with the color handling. Technically, it is not supposed to have this quality. If you talk to an engineer, he will tell you that DV doesn't reproduce colors as well as DVCPRO or BetaCam. But, especially with Canon's color pixel-shift technology, I feel the colors look fine. It also depends on what you compare the image to. I've noticed that if you mix footage, people really notice, they're more aware of the color differences. If you shoot everything from a single format and the same type of camera, people aren't usually aware of the image differences."

Using DV for Broadcast Work

The miniDV camera has proven versatile for a wide range of applications. "I've used it for everything," Victor said, "from music videos to surf videos to TV reports that are broadcast on European television. I just got back from Evian, France, where the woman's golf tournament, the Evian Masters, took place. The report that I did there was shot on DV."

"Most broadcasters prefer DVCAM," Victor acknowledges. "The report that I did, my cameraman was using a Sony PD-150. But, since I don't have any DVCAM machines in my editing suite, we just had him shoot it in miniDV (which you can specify on the Sony PD-150)."

Victor notes that, "There is virtually no difference between DVCAM and miniDV formats—they use the same compression. The DVCAM heads spin faster than miniDV and the tape itself moves faster, which, in theory, means that you get fewer dropped frames. Also, the audio is locked in DVCAM to minimize audio drift." MiniDV and the small DVCAM tapes can be interchanged; DVCAM tapes are generally considered to be more robust.

"Sometimes," Victor said, "if a project is very important, I will use the small DVCAM tapes instead of miniDV tapes."

Many moviemakers get attached to particular software that meets their needs and Victor is no exception. "I have a Macintosh G4 machine," he said, "and I am running an outdated software program called *Cinestream*. It was originally made by Radius and called *EditDV*. Radius sold it to Digital Origins and they sold it to Media 100, who changed the name to *Cinestream*. *Cinestream* recently sold the program to Discreet and Discreet just stopped developing it. As far as DV editing is concerned, it is the most intuitive and the fastest editing program out here. If I work with mixed media, I will work with *Final Cut Pro*, if I work exclusively with DV format, I usually go back and use *Cinestream*."

Victor first moved to France (considered the surf capital of Europe) to further his efforts with H3O TV after he had given up the notion of attending the film academy in Vienna. Recognizing that the film academy credentials can be a helpful entry into the competitive moviemaking industry, he still doesn't regret having taken a more direct path into the business. "I just jumped into it," Victor said. "I am just a self learner. A lot of that comes from my experience hanging out in libraries, so I buy books and learn from them. Now, with the Internet, I have access to this enormous library of information. I read and then I learn from my mistakes."

Overcoming the Bias Against DV

The traditionalist film community in Paris has trifling respect for works produced using the DV format. As filmmakers who are often subsidized by the government for their creative work, as a community they are less interested in creating films that have box office appeal and more interesting in exploring human dramas and psychological conflicts. Because the art form is steeped in tradition and married to decades of long-standing convention, DV enjoys little respect from so-called serious filmmakers. From the time that Victor first started working with DV tools, he was determined to show that with a professional approach DV can deliver surprisingly strong imagery and that it can be used effectively as a story-telling medium by anyone willing to master the craft.

Victor has worked with different film and videotape formats, including 16mm, Super 16mm, Sony BetaCam, DV, Hi-8 and 8mm, as well as working on a Super 16mm production that was blown up to 35mm.

"I still see people looking down on DV," Victor said. "They are very stuck with tradition and tradition says that film is what you take seriously. Almost all projects are financed by the government through the CNC and so there is a healthy, thriving film community, but it is not based on people being creative or taking risks—it is very bureaucratic."

Part of the difficulty of getting financing for a creative moviemaking project is overcoming the entrenched attitudes that prevail in the French film atmosphere. "The CNC has this view of what French film should be," Victor said. "If you work within that envelope, you are OK. It usually favors introspective works, based on people and situations and emotions. It is hard for me to describe. You just have to watch some French films to understand."

Nonetheless, some of the edgier creators are breaking the strictures of the form and moved in other directions. "You have Luc Besson, who did *Nikita* and the *Fifth Element*, for example. He is one of the few (and he is really not liked by the French film industry) who has broken away and said, 'You know what, I'm going to make something that is French, but appeals to a greater audience.' He is one of the few who actually makes films that work at the box office. *Amélie Poulain* was another film that works at the box office. It is a French film, but it is not very typical of French films."

Finding a Balance and Gaining Acceptance

"I enjoy being in Paris, because of this cinema auteur atmosphere, where the creative process is very important. When you are working on a film, it is considered more of a work of art that has nothing to do with business. Storytelling in film tends to be more narrative, more about what is going on inside of the characters. Coming from an American movie background, it is more about the action and what is happening around the characters. I like more of a balance. But I do like the fact that here you can shoot a film that reflects the director's vision, whatever that vision may be."

Given the somewhat insular nature of the French film community, Victor maintains a fairly independent stance in his working environment. "I work with technicians. I work with good sound engineers and good directors of photography, but otherwise I feel pretty much on my own here," he said.

In the quest for support and recognition, Victor has circulated his short film to production companies and producers. "I another short film that I have written," he said, "called *The Tao of Surf*. It is about a young surfer who learns important lessons of life through nature, the ocean, waves, and surfing. I also wrote a feature film, it is called *Digital New World*. It combines extreme sports with virtual reality and it all takes place in the future. Somehow, this doesn't fit into the typical French mold. Luc Besson's production company likes the short film, *The Tao of Surf*, but that is why I use him as an example of a company that doesn't fit into the French structure. I sent them the short film and they are considering it, whereas everybody else just doesn't want to hear about it."

Making the Movie

Tinged with martial arts mystique, *The Chinese Shoes* relates the story of a young boy, Lao, whose attraction to a girl in the neighborhood leads him into a humiliating encounter with a local bully. Lao finds a pair of Chinese shoes in an antique shop and dons them in the hopes of gaining the power of a kung fu master. The seed for this idea came from Victor's martial arts instructor in Paris, who described two fights that he was drawn into when he was wearing Chinese shoes. "Never wear Chinese shoes," his instructor warned ominously. The idea percolated in Victor's brain for a time until he decided to fill in the details of the story in a movie. The protagonist in the movie discovers that the misuse of power can be devastating and learns a lesson in the process, which causes him to return the shoes.

"I am treating my short film as if it is a Hollywood blockbuster in the sense that I believe in promoting it," Victor said. "I believe that I shot it as well as I could. Everywhere along the line, I am treating it as if it were a big budget production. This comes from my business background. I have to take a step back. It is my film, but it is also product that needs to be marketed and promoted. I found a niche (the fact that I shot it using the Mini-35 adapter and used professional cinematography techniques), based on this point of interest in the film. I am going to be targeting Asian-themed media afterwards and maybe martial-arts themed media. But the first marketing step was based on the technical aspects of this short film."

Freedom from Film

The attraction of the DV format to many new moviemakers is the freedom from the tyranny of film—both the expense associated with purchasing and processing film, as well as the greater difficulty in assessing daily results of a project and editing the content. Inexpensive non-linear editors and video that can be monitored and corrected at the source lead to more experimentation and more precise results in many cases. "The reason that I pushed to do this in miniDV was because I was making surf videos at the same time as I was reading a magazine called *Res*. *Res* was really talking about the art of DV filmmaking, about how DV was such a liberating format. I just loved the idea. I want to liberate kids, or anybody who is like me when I was a kid and wanted to make films, but I didn't know where to start. You can just buy a inexpensive camera to start and with the DV format, you can make something that is respectable. You just have to free your mind. Once you learn the techniques, it can lead in any other format. It is liberating in the sense that you can learn the techniques, and you don't have to worry about buying these expensive rolls of film. You can concentrate on making your movie and making it in the best way possible."

Victor also sees advantages on the distribution end of the process where the need to duplicate expense reels of 35mm film is eliminated. With more theaters adopting digital projection systems, such as Landmark theaters as discussed in Chapter 18, distribution of movies can be handled by means of broadband networks or even DVDs. Anything that makes it easier for an independent moviemaker to get work shown on the screen is a positive factor in Victor's judgment.

"When movies eventually go digital," Victor said, "it will be even more accessible for the independent filmmaker, because you won't need to have big rolls of film that you need to produce and ship around to theaters. Jean-Marc Barr who was making Dogme films, did all his films with on DV." Dogme films are the creation of Lars Von Trier and Thomas Vinterberg. In 1995, they set up the rules they considered valid for creating uncorrupted movies. As part of this aesthetic, the DV content is sometimes degraded to give it a more harsh video look.

A good example of DV's potential, Victor believes, is Danny Boyle's work, *28 Days Later* (*www.28dayslaterthemove.com*). Victor said, "It is a feature shot with a Canon XL1 while maintaining professional cinema techniques."

The accessibility of DV is also its Achilles heel, Victor believes. "One thing that gives DV a bad name is that all of a sudden it has become so accessible. People think that being accessible means that you can do whatever you want without learning the craft. But, you have to take DV and treat it with respect—then you'll get quality images."

Struggling to Complete a Film

A classic tale of the obstacles encountered making a film, Terry Gilliam's *Lost in LaMancha* (*www.smart.co.uk/lostinlamancha/*) should be mandatory viewing for anyone interested in independent filmmaking. "Terry Gilliam didn't have enough money to make his film," Victor said, "and there wasn't any margin for going over budget and over schedule. So, of course, that is exactly what happened. They just had a couple of days of hard rain and then one of the actors, the person who was going to play Don Quixote, came down with a prostate infection. That sent the whole project over schedule and the budget just buckled. They had to fold up production."

Like Terry Gilliam, Victor struggled with actor not showing up and also dealt with unpredictable weather and other kinds of unexpected contingencies that cause directors to lose sleep in the middle of production.

"The weather and the actors vexed me," Victor said. "The main actor couldn't get off of work, because he is only a part-time actor. So, we were shooting around his working schedule. Luckily, he worked near the neighborhood that we were filming, so he would sneak out of work, come and do his scene. When we were setting up for the next shot, he would go back to work. The actor who was going to play the role of the shopkeeper just decided not to show up. I called him later and asked him, 'why didn't you show up?' He says, 'I just changed my mind. Everybody who knows me knows that I am like that.' Well, I had rented the shop. I had paid the employee of the shop to be there to supervise us so we didn't break anything. I had already put out all the money for the equipment and gotten all the crew together. I had rented the equipment for exactly five days. So, if a day of shooting fell through, that meant I would have to re-rent the place and add another day to the shooting schedule. I could have had a nervous breakdown right then and there. The whole crew was just looking at me like, 'What are we going to do now?'"

Victor scoured the neighborhood for someone to fill in for the missing actor. Somehow, he knew he would find somebody. As Victor describes it, "I found an Asian art gallery owner who said he had acting experience. I didn't even bother to check. I just asked him, 'Do you want to do it?' He said, 'Yeah, sure.' So brought him into the production."

Throughout the production, obstacles kept popping up one after another. "There were so many things that came up that I thought were going to just kill the project," Victor said, "but there was always an exit. When we wanted to do the fight scene, which was an outdoor scene, it was always cloudy and then it just started raining. I eventually had to postpone the shoot for a month."

A clothing company that had agreed to sponsor the film, Netherlands-based G-sus, was late in shipping the clothing to be worn by the main character and other cast members. "If you look at the characters in the film," Victor said, "other than the shopkeeper and Lao, the main character, all of the other characters are dressed in G-sus clothes. The late delivery didn't affect me too much, because the scenes that we were able to shoot didn't involve any of the other characters; it was only just Lao and the shopkeeper. I had to postpone until September to shoot the fight scenes and by then the clothes had arrived."

A Compact Crew

Victor worked with a compact, but experienced crew consisting of a cinematographer, a sound engineer, boom operator, grip, gaffer, script girl, and behind-the-scenes videographer. Curiously, the cinematographer, Pascale Marin, was someone that Victor met by chance in Luxembourg park in Paris. Pascale was practicing martial arts moves. Victor approached her about starring in his movie. It turned out that she was a graduate of the well-respected Louis Lumière School of Cinema and after some discussion, Pascale agreed to oversee the technical aspects of the video work in the movie. She worked closely with Victor on taking maximum advantage of the Cooke lenses and Mini 35 adapter to achieve optimal images and the desired effects.

Figure 15 - 1 **Filming begins (photo credit: Katrin Gunterhausen)**

A production manager would have been a helpful addition to the crew, Victor reflected. "I was completely swamped with responsibilities and a production manager could have taken a lot of the load off my shoulders so that I could concentrate on making my movie without worrying about the logistics of the shoot."

Sound was routed directly to the Canon XL1 rather than to an accessory DAT recorder, a factor which saved a significant amount of time in post-production. The sound engineer monitored and adjusted the audio being miked by the boom operator. An audio post-production house added sound effects, but did nothing else to the audio content.

Victor licensed a 70's song, *Kung Fu Fighting*, for a musical backdrop, negotiating a limited license that allows use of the music for film festival purposes. Once the movie goes into distribution, Victor will pay additional rights on the song.

Figure 15 - 2 **Lao flips bully (photo credit: Katrin Gunterhausen)**

Audio Work in Vancouver

Because of the Internet, Victor was able to accomplish many of the audio post-production tasks remotely. As Victor explains it, "I had nine tracks of audio and I just exported them as AIFF files. Bundled them into a folder and compressed the folder with Stuff-It. I uploaded this file to the Vancouver-based sound mixer's FTP server. He then did the mixing and he would send me an MP3 file of each version of his work."

The synchronization of the audio and the video was handled by inserting a one-frame beep at the same time as there is a single-frame of color bar, which occur three times. "If the beep and the color bar were synchronized," Victor said, "that meant that the audio and video tracks would be synchronized."

The sound mixer, David Raines, sent him successive version of the files, Victor would sync them with the video and send back comments for handling the next round of changes. After several rounds of changes, Victor had the audio tracks exactly the way that he wanted them.

The process was facilitated by the use of a DSL Internet connection. Although the initial file sent to David consumed 400MB, the interim MP3 files during the reviews and revisions only required 12MB—the storage size of a 15-minute MP3 file. The download of this file at DSL transfer rates only required about 5 minutes. The audio mixing was done using Pro Tools.

Victor gained the benefit of less expensive audio mixing rates in Vancouver, plus the expertise of David Raines (*www.screensoundinternational.com*), who has worked fairly extensively with DV and DVD sound tracks. "I picked David Raines," Victor said, "because of his experience with 5.1 surround sound. He was offering a price that was better than anybody could offer me in Paris, as well as outside of Paris. And, he was familiar with working over the Internet, so, that clinched the deal."

The initial meeting with David occurred through a Web site geared to bringing people in the audio-visual industry together: *www.mandy.com*.

Creative Financing for Independents

Independent filmmakers without deep pockets often have to get creative about how they handle the expenses that accompany any creative project. Victor relied on ideas he had developed during years of low-budget production work as well as some tips that he gained from a book called *Film and Video Financing* by Michael Wiese. "That book gave me a lot of ideas about how to make *The Chinese Shoes*. The author talks about different approaches, like the deal I cut with the clothing sponsors. Product placement is one key idea—I used that technique to help finance the film."

Catering on a Shoestring

If you're making an independent movie, cast and crew need to eat. Catering expenses can be considerable, even on a short five-day shooting schedule. Thinking in terms of the product placement techniques Victor read about in *Film and Video Financing,* he approached a number of restaurants in the Paris area hoping to exchange exposure in his film in return for catering services. "First I contacted the big chains, like Pizza Hut and McDonalds, but I ended up with a small falafel restaurant near the Bastille. I told them that I would include a shot of their restaurant in the film in exchange for them providing the catering. That saved me a good deal of money. At least I didn't have to feed at a dozen people per day."

Victor worked the product placement into the movie sequence in a natural way. "There is a scene when the main character is prancing down the street, showing off his new Kung Fu skills," he said. "He prances around right in front of the falafel restaurant, *L'As Du Falaffel*. There are a couple of patrons who come out and he startles them."

Victor felt that the restaurant primarily wanted to support an artist in their own way, but that the additional exposure for the restaurant was also a plus, in their estimation. They also received a mention in the credits for the movie.

Shooting with a Mini 35 Adapter

The Canon XL1 is one of the few prosumer DV camcorders that lets a filmmaker interchange different lenses. This feature turned out to be a major benefit for Victor to accomplish the precise effects that he wanted to create in his movie. Once again, some creative bargaining was involved, since the cost of the Cooke lenses and Mini 35 adapter that Victor wanted to use exceeded the purchase price of the Canon XL1 camcorder.

"To get the lenses and to get the use of the Mini 35 rig," Victor said, "I basically suggested to them that they needed footage to show off their new system. So, the deal was that I would provide them with footage they could show in the trade shows."

Technical Aspects

The Chinese Shoes was shot in widescreen 16:9 aspect ratio on the Canon XL1, which Victor feels works very effectively on this camcorder despite the 3:4 image ratio of the camera's three charge-coupled devices (CCDs). However, the standard Canon 16x zoom lens didn't offer a wide enough angle for some of the shots that Victor had planned, including some crucial interior shots in the antique shop. The widest angle on the zoom was the equivalent of a 38mm lens (in 35mm terms) and even the standard

Canon 3x wide angle lens (with the equivalent to a 24 to 72mm range) wasn't sufficient to handle the scene as envisioned.

Several of the shots planned also required very specific focus and depth of field settings, something that is especially difficult to achieve with the standard Canon lens and the servo-controlled zoom ring. Even at maximum aperture settings the depth of field was not adequate and using rack focusing techniques in a shot proved extremely frustrating due to the lack of markings on the lens barrel. Certain adapters, such as Canon's EF lens adapter system, could open the camera up to a wide range of EOS photographic lenses, but at a cost. The adapter produced a 7.2x image magnification because of the differences between the camcorder's image plane (measuring, in this case, 1/3-inch, the size of a CCD) as compared to the conventional image plane in a 35mm camera. This adapter also would not produce a solution to the depth-of-field problem.

A solution was found in the Mini 35 digital lens adapter, manufactured by P + S Technik, a versatile system that could accommodate a set of Cooke S4 Optics while preserving the depth of field and focal lengths associated with a 35mm camera. An added bonus in this situation is that Victor was also able to attach a follow focus device, which enables the lens focus to be controlled precisely when the distance changes between the camera and the central point of interest. This added one more professional element to the range of options available during the shooting.

Victor would like to be able to use this system for future projects, but he said, "I'm not planning on investing in the equipment. The Mini 35 system costs about six thousand dollars. The lenses themselves are also in the thousands of dollars. This is something that you would normally rent; at least something that I would rent, because I don't have that much use for it. Although I would like to shoot everything with this system, right now it is not worth the investment."

The arrangement with the rental agency worked out well for everyone involved. Victor only had to pay the insurance on the Mini 35 system and Cooke optics and the P + S Technik obtained some video footage demonstrating the considerable capabilities of their system. The insurance for this equipment turned out to be about 10 percent of what the actual rental price would have been, somewhere in the range of three hundred dollars.

Project and Distribution Costs

In the spirit of independent filmmakers everywhere, Victor negotiated every chance that he could to get better prices and terms for this project. Obtaining the use of the Mini 35 adapter and Cookes optics was a major coup that had a substantial effect on the results of the movie. The clothing provided by G-sus was used as a token payment to cast and crew. The falafel restaurant handled the catering. Out of the pocket expenses were primarily the lighting and sound equipment and the rental of a truck to transport all the gear. Victor estimates that the actual production costs came to somewhere under $1000. Post-production costs, including the audio mixing, will total somewhere around another $1000. Shooting the same short film in 16mm or 35mm format would cost several times this amount. However, Victor is considering transferring the digital video content to 35mm film, which would entail conversion costs of somewhere around $5000. The alternative is to look for film festivals and other venues that are more open to non-film formats, such as miniDV or DVD submissions.

Gearing Up Promotion

If all goes well, *The Chinese Shoes* will be hitting the film festival circuit at the end of 2003. Victor has already begun promoting the film through a Web site and magazine articles that have been published in *MovieMaker* and a French periodical.

The five-thousand dollar conversion cost to 35mm is daunting. "I have to decide if the film festivals that I am entering are worth the investment of having it blown up to 35mm. But the thing you have to understand is that I shot this 15-minute film for about $2000. If I had done this on film, it would have been easily three times or four times that. So, what is five thousand dollars if you look at the big picture? If you are nickel and diming your film, it is just so out of proportion, you tend to lose track. It is still a healthy sum of money," Victor reflected.

Victor is still investigating film festivals that encourage DV entries, hoping to gain some additional funds to bootstrap his production and perhaps enter the film in other festivals.

Bringing in an Editor

As another concession to mainstream professional practices, Victor brought an editor into the project who normally works on cinema auteur films to help shape the story presentation and pacing. "I chose him," Victor said, "because I come from more of an action background. I knew that I would need somebody to work with to make the movie really tell a story." This editor also writes and directs film projects, so Victor felt he could rely on the additional expertise on sharpening the storyline.

The editor worked from a rough cut of the movie and listened to Victor describe what he wanted to achieve with the story. The editor's work had a significant effect on the tone and character of the movie.

"The editor's influence was especially evident in a scene in the antique shop, where we didn't have a lot of time to shoot," Victor said, "There was a missing actor at one point and the main actor was taking off from work. That particular scene was saved by the editor and my director of photography. We started in the morning. We ended up filming until three or four o'clock the next morning. Pascale had to transform the lighting in a night scene and a day scene. We ended up shooting with very poor light conditions, but she really did some magic with the lighting. She kept the scene looking like it was shot at the same time the whole day and into the beginning of the evening. It was something where I would have just given up, but she really worked with just very limited lighting to create the illusion that we were shooting at a certain time of day, that everything was happening in one specific moment in time."

Victor has strong ideas about his work, but he also recognizes when a project can be improved by bringing in additional expertise to add depth and dimension to a movie. As he describes it, "I had very positive experiences with my cinematographer, editor, sound person and sound mixer. I enjoy working with other people and, generally, I am pretty laid back. I choose people to work with because I respect what they do. So, I should listen to them. I have my own vision of how my film should look and I will have the last say. But I am not so fixed on the notion that I'm the director and it is my film. I see so many talented people and the reason that I am working with them is because they are talented."

Getting Into the Festival Circuit

The festival circuit represents an avenue by which an independent moviemaker can achieve recognition, but so far the festivals that accept native DV format for submissions are few and far between. Although eventually Victor would like to get *The Chinese Shoes* transferred to 35mm film for the moment he is seeking festivals that accept DV format or BetaCam. "The San Francisco International Asian-American film festival will accept BetaCam or 16mm or 35mm," Victor said. "So, I am shooting for 35mm, but if I can't afford it (since it is going to cost about $5000 US), I can always give it to them on BetaCam."

The quality differences between DV and BetaCam are somewhat equalized by the fact that some degradation of BetaCam content occurs once it is transferred into a digital editing suite. Victor, who has worked extensively in both formats, explains it in these terms, "When BetaCam images are originally captured on tape, we can say that the quality is better than DV. But, when you transfer out of the master tape and into the editing suite, it generally gets transferred out in analog format, which will, in my opinion, degrade the signal a little bit. Then, you put the video into a computer, which either applies one form of compression to it or captures it without any compression (which takes a lot of disk space). It captures the content in its degraded format. When you edit it, there is more loss in the edit. And when you output it again, you usually output it using analog cables, back out onto the BetaCam deck."

"With DV," Victor continued, "if you use FireWire, when you transfer the rush into your editing suite, there is no loss. When you edit, there is no loss. When you output it back onto DV, there is no loss. That is the magic of working with DV. Your finished product is not going to have any image degradation. It is an exact copy of what you originally shot. Unless you have adjusted the colors or added special effects or things like that."

The Canon XL1 includes a Frame mode of shooting, which is similar to 24p format, but for filmmakers intending to have their work transferred to 35mm, you should check with the lab you intend to work with first to see what their requirements are.

"Frame mode is an aesthetic choice," Victor said. "I shot *The Chinese Shoes* with the goal of transferring it to 35mm. I talked to the lab that I decided would do the 35mm transfer and asked them how they preferred the DV. They wanted it interlaced, so I shot it interlaced. With their process, they interpolate the frames themselves and they get a better resolution than if I shot it in Frame mode, which isn't that bad. Frame mode, when I shoot in PAL, it is 25 fps, which is very close to the 24 fps of film. The Frame mode on the Canon is not supposed to be a true progressive scan frame, so there is a drop in quality. It is an aesthetic choice. You have to decide if you want your video to look like film on a video screen or if you want it to look like film on a real screen. If you are never going to put the movie on film, it would behoove you to maybe shoot in progressive mode or frame mode. You can then go back and either use After Effects or any of the other programs that add that film look. Understanding that the video is never going to go on film, but you want it to look like film. If you want it to go on film, you want to keep the resolution at the highest possible level. Contact the lab that is going to do the tape-to-film transfer and ask them how they want the video."

The benefits of entering the film festival circuit can be intangible and elusive, but it is one means of attracting notice and establishing credentials in a crowded field. "I don't think people land major films from winning in festivals," Victor said. "Because of my background, I don't have any formal schooling or degrees in filmmaking. My experience is based on surf and sports productions. I don't know what the situation is like in the States, but I know that when I came to Paris, I could have said that I was making movies with Martians. They would just look at what I had done blankly. They might be intrigued by surfing, but they wouldn't take it seriously. So, being on the film festival circuit could give me the credibility that I need to launch other projects with other people. If I want to work by myself and self produce my own works, which is not out of the question, it is still good to be able to talk with sponsors and financiers. Something to consider: there are a lot of people who say 'I want to make a movie,' but they don't actually follow through. If you have a product that is completed and that is being shown, then you have gone the full loop. People will take you more seriously."

Putting The Chinese Shoes on DVD

"I want *The Chinese Shoes* on DVD, but I don't yet see a market for a 15-minute short film on DVD," Victor said. "I don't know who would buy something like that. I want it on DVD, but it needs to be packaged with other short films, along in the same genre. If a distributor came up and said, 'Listen, we loved your film and we want to put it on a DVD with other short films, I would jump at the opportunity.' I already did a DVD last winter and I'm in pre-production for the next one. It is an extreme sports movie that actually tours movie theaters, something like the Warren Miller film tour. Last year, the movie theater distributor wanted us to do this project on DVD, so that it could be projected from DVD, just as Warren Miller is doing in the states. This year we might just transfer it to 35mm, because it is easier to distribute to a wider ranger of theaters."

"I am excited about the new HD DV format," Victor said. "I am looking at it as a way of capturing video and then converting it to DV and editing it and coming out with a DV video that has much higher resolution."

16

Case Study: Distributing Content on DVD

Calvin College spans 370 acres in western Michigan with a history and ethical compass inextricably linked to the Christian Reformed Church. Over the years, a vast collection of audio and video material has been amassed, covering classes, training seminars, special events, and other types of instructional material. A quick tour of the college's archives provides a stark reminder of how many media formats have seen the light of day over the past 30 years. The shelves contain reel-to-reel black and white video, 3/4-inch videotape, VHS cartridges, as well as audio cassettes and other media types, including some very old "wire" recordings.

With earlier generation video playback equipment becoming increasingly difficult to maintain, the staff of the Instructional Resources Center had a strong motivation to transfer audio and video materials to a more durable and long-lived archiving medium. To this end, Director Randy Nieuwsma took up the challenge to convert the library of past media content to a more viable format for the future: DVD. The same scenario no doubt exists for countless universities, news organizations, libraries, research centers, historical organizations, and others—the process of archiving video content to DVD can provide a measure of longevity to important video material, as well as providing a mechanism for making content more accessible through a cohesive menu structure and intuitive navigation system.

Desktop Duplicators in Mainstream Use

One of the more noteworthy advances in the DVD technology realm is the prevalence of affordable desktop DVD duplicators in mainstream applications. No longer the purview of well-heeled corporations, desktop duplicators make it possible to set up on inhouse video publishing operation, inexpensively producing DVD discs in all formats for little more than the cost of the recordable discs. For archiving or duplication tasks limited to two or three hundred discs a week, a desktop DVD duplicator can serve as the centerpiece of a small-scale optical disc publishing operation.

After researching and comparing the models on the market, Randy Nieuwsma purchased an ElitePro2 CD/DVD duplicator equipped with two Pioneer AO4 recorders. Randy had more than one goal in mind for the duplicator—beyond archiving the old audio and video collections on CD and DVD, he wanted to establish a foundation for DVD duplication within his department. The archiving process is well underway, and the DVD production capabilities are being used as a revenue-producing channel.

Current-generation desktop duplicators can typically encompass the full range of CD and DVD formats. The most significant division in capabilities is whether the DVD recorder is set up for DVD-R or DVD+R formats. The compatibility with set-top players is roughly equivalent at press time (approximately 80 to 90 percent with a slight edge in favor of the earlier DVD-R format). But, the prevailing trend is towards more extensive compatibility so that the newer players are able to play discs produced under each of the existing formats. This situation should continue to improve, as it did for the multitude of CD formats which can now be played back with universal compatibility in the vast majority of computer drives and set-top players.

Many DVD duplicators also simplify copy processes with flexible options for processing a succession of masters. In other words, you can initiate an overnight operation by stacking the feed bin with an appropriate assortment of masters and blank discs. This feature, as Randy described it, works this way, "Let's say you have three master discs and you want five copies of the first one, one copy of the second one, and ten copies of the third one. You can create a sandwich in the input bin. To duplicate five copies of one, you put five blanks down and put the master on top. If you want one copy of the next, you put one blank down and put the master on top. And so on. In this mode, when the duplicator receives a disc that has material on it, it says, 'Oh, I'm going to make an image of this and write it to as many blank discs as follow.' That is a very cool and useful feature."

Wrestling With Emerging Technologies

The process of setting up and implementing DVD duplication for the Instructional Resources Center was not as simple as Randy had hoped. As with the introduction of CD recorders, the process for early adopters of DVD duplication has been a challenging one, made more difficult by the preliminary state of the documentation for duplicators.

"As the Director of the Instructional Resources Center at Calvin College," Randy said, "I work with a number of departments that handle media. That includes Audio-Visual, Video Production, Publishing Services, and Instructional Graphics. We've already done a good deal of VHS duplication and audio cassette duplication and I figured, 'If I don't jump in here and get involved with DVD duplication, somebody else is going to beat me to it.' Media duplication is what we do. I jumped in early, knowing that I was paying a bit more for the equipment and knowing that the technology was not quite as mature as I would like it to be."

The initial plan was to turn much of the responsibility over to the students, but, in reality, that wasn't possible as quirks and complexities made the process difficult to manage. "We use a lot of students to do the work in this department," Randy said. "I had thought that once I figured out the duplication system, I would write up some procedures and then turn the process over to the students. I'm now realizing that I won't be able to do that for some time to come."

"Some of the complexity is necessitated by the increased level of services provided," Randy said. "For example, each audio and video program requires the creation of at least one unique label. Programs that are available in more than one format, such as audiocassette and audio CD or VHS and DVD, require two labels to be created. Even though those labels are made from pre-existing templates, you need a bit of an eye for design to handle some variables, like very long titles or listing of multiple speakers, and so on. Then, those files must be stored in a system that allows them to be easily located and printed. Obviously, as the sheer number of titles increases, the problem of securely storing and accessing those masters increases. On top of all that, there may be copyright restrictions on the distribution of the title. The result is more complexity than what I am comfortable assigning to an inexperienced student."

Unattended Operations

One of the major benefits of the latest generation of DVD duplicators is the time-saving and ease-of-use afforded by unattended operations. Many duplicators combine the optical recording and label printing functions in a consecutive sequence, usually relying on a simple form of robotics that transfers recorded discs to an inkjet or dye sublimation printer. Once the label has been printed, the robotics remove the disc and stack it on a spool of completed discs. This type of unattended operation, of course, requires a high degree of reliability from both the operators and the equipment. No one wants to return after an overnight disc duplication session to find the equipment locked up from a hardware or software error.

Randy frequently takes advantage of the unattended operation features of the ElitePro2 duplicator. "You can have unattended 'burn discs' and you can also have unattended 'burn and print.' Or, you can print the label later. If I have a lot of discs to run, I let the program print the label. To print the label, you have to save label image as a PRN file (a print image file) and then define the path of that label and the program prints it."

Database Searches

The historical progression of diverse media formats and an uncoordinated system of cataloging content led to increasing difficulties in locating specific materials in the instructional resources center. The natural answer to this challenge once again relied on the computer. "I had an epiphany about five years ago," Randy said. "The solution to my problem is a database. I created a database that has information about every event that we've recorded—including classes, seminars, presentations, lectures, and so on. The database includes the audio information, the video information, and even the print information. All of this is on our internal network. I'm not the only person who works with the media. Anybody who works with it can enter titles and then it automatically reports from the database to a Web page where you can conduct a full search. The Web page lists titles alphabetically, performers and speakers alphabetically, and series alphabetically."

"The public also has access to the database and can run searches," Randy continued. For example, someone might want to track down the works by certain authors that they know or find a lecture from *The January Series*. They can look up an item by title or series and find the item number used by the bookstore. The bookstore orders get routed to me. I make the

DVDs on the duplicator and send them back to the bookstore. The bookstore sends them off to the public."

Going Digital

When the DeVos Communication Center at the college opened in late 2002, a prodigious collection of media content from the old facility was creating storage problems and gradually deteriorating. Rather than simply moving the entire collection to the new facility, the Communication Arts and Sciences department began discussing other ways to handle the content. "They called me," Randy said, "and told me they had video dating back to 1979 of all of their major shows (they do three plays a year). They were concerned that the archives took up too much space and wanted to convert to a digital format, if possible. So, I made a DVD of every play that they had for long-term archiving."

Using a Pioneer DVR-7000 DVD recorder, Randy completed the process of digitizing the complete collection to DVD. "We checked them all," Randy said. "They all worked." The originals were disposed of and the new collection was transferred to a glass storage case, where discs could be loaned out to students or faculty on request. "One day," Randy continued, "one of the professors called and asked, 'What happens if we lose one of these—if we loan one to somebody and it doesn't come back?' I explained that they would be out of luck, just as if they had given someone an original videotape."

As a result of the conversation, a dual master system was established. One set of discs resides in the safety of the college archives and another, created with the help of the DVD duplicator, resides in the CAS Department archives. "For the professor," Randy said, "it's the best of both worlds. He has a complete set of videos that he can use and has the confidence they won't ever be irretrievably lost. We can create as many copies as necessary whenever they're needed and each copy looks as good as the master. This application of a DVD duplicator made a lot of sense for us."

DVD Archiving Possibilities

Many different organizations have stockpiles of media content similar to what Calvin College is contending with. If you travel back a span of forty or fifty years, you can view a steady progression of media types, each one incompatible with the last, and most of them based on analog data storage techniques. The reliance on analog storage, of course, ensures there will be a generational loss from copy to copy, with additional noise and signal degradation introduced each generation.

As an archiving medium, DVD avoids generational loss during copying; each bit recorded on the master disc is faithfully recorded on the archive copy. The 100th copy of a DVD master exhibits virtually the same quality as the original, assuming that the media used for the copy operation is good quality and that whatever utility is used to perform the copying completes a verification at the end of the operation to ensure the integrity of the copied disc. Archivists, historians, media librarians, and others who want to preserve rapidly aging content stored on magnetic media or film can take advantage of the archival possibilities offered by DVD.

With DVD media lifespans estimated at greater than 100 years, the critical issue becomes whether there will be playback equipment available somewhere down the road when someone wants to read the archive. Odds are that at some interim point, digitized content will be transferred to whatever form of media becomes the successor to DVD.

In this regard, Randy commented, "I think we are close to considering the network to be the next digital format. The distinction between location and format dissolves when we realize that the network folks are charged with archiving, protecting, and providing access to all digital content *forever*. The actual hardware and systems they use to do this are transparent to users. They transfer the existing content whenever they upgrade their storage devices. From my perspective, once the content is digital and on the network, it will be available forever. I can retrieve it and record to whatever format I need."

Mastering Everything to DVD

In September 2002, Randy made the decision to use DVD discs as masters for the full range of video content produced for the college. Both the archiving of the earlier content and the storage of newly developed content would be written to DVD and duplicated, as needed, for distribution to other departments or for sale to the public through the bookstore. The decision had wide-ranging implications that affected nearly all of the production tasks performed by the department, but in many ways the production process has been significantly simplified by the DVD storage and duplication.

As Randy describes the transition, "Last September was the dividing line, after which we planned to master everything to DVD. We do a lot of video production in this group and another department does a good deal of audio work, primarily covering events and conferences. We do a fair amount of internal production and we also produce a panel discussion that we air on the local PBS affiliate. In our video production department, I have two people who shoot video on miniDV. We also use DVCPRO, which is our best format with the highest quality. We do have a portable deck that we generally take along with our best DVCPRO camera to capture events, such as when an important figure is coming to campus to do a lecture presentation."

Producing The January Series

One of the high points in the video production schedule is *The January Series*. "*The January Series*," Randy said, "is a nationally known series produced, surprisingly, by this little college in Michigan [Calvin College]. We have won the national cup three times, beating out Harvard, Yale, and Michigan State. Not bad for a little school in Michigan."

The January Series draws a good deal of positive attention to Calvin College through a series of lectures and discussions that explore a wide range of eclectic topics. Subjects span different points of view from many different fields: sports, the arts, popular culture, religious philosphy, ethics, and so on. Speakers also come from a variety of fields, including academia, music, art, journalism, and literature. For example, The January Series in 2000 included *Sports Illustrated* contributing editor Frank Deford and Terry Gross of National Public Radio.

In 1999, The January Series brought accolades to Calvin College in the form of the Silver Bowl, awarded by the International Platform Association, an organization devoted to excellence in college lectures that was founded by Daniel Webster. Calvin College has won this award three times over the span of a decade.

Students and staff members at Calvin handle the video production work for the series, including camerawork and editing. Individual events from the series can be purchased on DVD through the campus bookstore. Randy oversees disc duplication as orders from the bookstore are forwarded to the Instructional Resources Center.

More Duplicator Uses

Besides the formal uses of the DVD duplicator in the instructional resources center, more casual uses generate good will for the department and highlight the capabilities of the technology. The academic year at Calvin is a 4-1-4 model. The two semesters are separated by a three-week long term in January called *Interim*, in which students take only one class.

Many of these intensive classes combine fields of study and are held off-campus.

A recent Interim turned students loose in Australia for an extended bicycle journey across parts of the country. When the group returned, Randy collected photographs and video content captured by the students and created a DVD documenting the adventure. After creating master discs, he designed a color label using a photograph of the students clad in biking pants at the edge of a sparkling Australian lake. Prepping the duplicator for the process, he started the operation and came back the next morning to unload a completed stack of CDs and DVDs with fresh, dry, inkjet-printed surfaces. At the post-Interim party, the CDs and DVDs were handed out to the students who expressed amazement and delight that their Interim expedition was immortalized on optical disc.

Improving Efficiency Randy is in the process of reorganizing the video duplication process to eliminate unnecessary steps. The new approach will integrate the video editing equipment and video duplication tools, setting up a direct connection to stream edited video content from the non-linear editor into the Pioneer DVR-7000 standalone DVD recorder. With the direct link to the NLE hardware, it will no longer be necessary to record a digital tape and then import the tape contents into the DVD recorder. This technique will allow Randy to effectively address the four major issues with which he contends:

- Creating a master recording in a durable format (DVD) that can be quickly distributed. For example, a DVD of each episode of *Inner Compass* is sent to the local PBS for broadcasting as soon as the editing has been completed.
- Shortening post-production time by using the DVR-7000 to digitize DVD's MPEG-2 format in real-time, as opposed to using non-real-time technologies.
- Gaining flexibility in the distribution formats. Copies of the master DVD can be quickly produced with the ElitePro2 duplicator and VHS tapes can generated using the traditional array of VHS recorders.
- Providing a long-term digital video storage format for convenient archiving.

As described in the next section, being able to produce very quick turn-arounds of *Inner Compass* episodes, from final edited version to broadcast-ready DVD, is a top priority.

Producing Inner Compass

The MPEG-2 video format that is central to DVD was originally devised as a television broadcast format. As such, it makes an excellent exchange medium for providing a television station with a completed show ready for broadcast. Calvin College takes advantage of this fact, using DVD to increase the quality of distributed episodes of their internal production of *Inner Compass* as delivered to the PBS station locally. Producer Jazmyne Fuentes shared her perspective on the process of producing individual episodes in the new DeVos Communication Center at the college.

Starting a Career in Video

Jazmyne Fuentes began her career as a video producer working for community access TV in Boston. From there, she got a job at WGBH working on educational print to go with their children's shows. "I had always been interested in TV production," Jazmyne said, "but that was how I finally got started, through those two jobs."

"When my husband got a job here in Grand Rapids," Jazmyne continued, "I looked around to see what kinds of jobs there might be at the college that had to do with video production. I found the video production unit and the Calvin TV show that had been started eight years earlier. The show was called *Calvin Forum* and it was primarily just a show for Calvin College and people in the local community, talking about different religious topics with a more narrow focus than we have now."

Choosing Topics and Finding Speakers

Whenever possible, Jazmyne tries to take advantage of the availability of people who are invited to the college to speak. The conferences and forums conducted at Calvin College attract a diverse collection of authors, philosophers, scholars, and other luminaries from around the world and many of these speakers become prime candidates for appearing as interviewees on *Inner Compass.*

"I usually keep my finger on that pulse," Jazmyne said, "to see who is coming. If there is anything associated with a speaker that is issue oriented, that is my focus. I try to do anything that involves social justice, religious, or ethical issues. That is my first pool of people and ideas. Then,

I get ideas just listening to the news and hearing about issues that really make me mad. I might think, 'That is just really outrageous; somebody should do something about this. Maybe I should do a show about this.' It is a great outlet for whenever you feel that something needs to be done about an issue."

WGVU, the local PBS affiliate, varies the time slot when the program is broadcast, which has contributed to keeping the ratings fairly modest so far, but presently *Inner Compass* is appearing on Tuesdays at 6 p.m., a time slot which should help ensure larger audiences.

Jazmyne played a key role in gaining placement for *Inner Compass* by broadening the scope of topics covered and helping improve the production values of the segments. "The show used to be more of a Calvin College-oriented show on the higher education access channel of the local cable TV system. Grand Rapids Community College did most of the programming and we offered this one half hour a week. PAX TV approached us and asked if we wanted to trade some local programming for air time, because they had a mandate for more local programming. When I felt that we had gotten good enough (I waited a couple of years), my executive producer in Public Relations and I approached WGVU and asked them if they would consider airing the show. They liked it and they really liked the rating that they had gotten for the Calvin College-Hope College basketball game, which is a legendary regional rivalry. They knew there were a lot of Calvin fans in the area."

With the thousands of Calvin College supporters in the area, there is a fair degree of interest among the community members just because of the college affiliation for *Inner Compass*, but Jazmyne would like to continue enlarging the scope of issues and topics to reach a wider market. She is hoping for more widespread distribution as they gain experience and skills in the interview format.

Handling the Production Details

The video production crew shoots in Panasonic's DVCPRO format and editing takes place on an Avid editing suite. Students may participate in performing a rough cut of each episode and Jazmyne proofs the results, correcting where necessary to maintain the high standards necessary for broadcasting. Rob Prince also performs editing of the video content, which currently is outputted to DVCPRO tape for conversion to DVD. The process, however is being modified, as described in *Improving Efficiency* on page 224, so that the output of the editing suite can be used as direct input to the DVD recorder.

DVCPRO shares the essential compression and sampling techniques of the DV format while being associated with equipment designed for more professional uses, such as within the broadcast industry. DVCAM, a similar format enhancement designed by Sony, accomplishes similar goals, making the DV format acceptable for high-end uses.

The PBS station accepts a DVD for broadcast use, which is dubbed to 1-inch videotape to correspond with their broadcast equipment. Previously, the program was submitted on 3/4-inch videotape but the conversion to 1-inch tape produced lower quality content than the transfer from DVD.

"The quality is definitely better," Jazmyne said. "We asked if we could start giving the masters on DVD and they worked it out the details on their end. Either way they dub to 1-inch tape, but the process is cleaner from DVD."

Logging

Jazmyne and Rob have worked out a set process for the video editing. "I digitize and Rob edits," Jazmyne said, "and then I proof. The process we've worked out is this: we go up to the Avid. I digitize the DVCPRO tape into the Avid and log it at the same time. I do a written record of the conversation in brief and then we take the preceding episode and make a copy of it in Avid and just work from that, changing all the titles as necessary."

Individual segments of *Inner Compass* run 26 minutes and 45 seconds, which is what the local PBS station requires. "It is not a stringent requirement, Jazmyne said. "We have gone a couple seconds over before, but we try to hit that nail right on the head because it helps them plan their promos."

Hosting Episodes

While Jazmyne chooses the topics and interviewees, she works closely with the hosts to brief them on the content areas of interest and to help develop relevant questions. "I find suitable guests," Jazmyne said, "and then come up with 40 or 50 initial questions. Once I narrow it down to the best 20, I give the host some written material to read at a meeting where we talk about the topic a little bit and go over the questions. I will usually present the questions that I have thought of and, if a host has any others that he or she would like to add, they can write them in on a document that I give them. We talk for about an hour about the topic and then they go and read for another hour or two on their own."

Quentin Schultz, who appears in a sample segment on the DVD bundled with this book, will host some of the upcoming episodes of *Inner Compass*. Shirley Hoogstra, who hosts the episode on the bundled DVD, will be alternating with Quentin and another host.

On the Set

The new set which has been constructed for *Inner Compass* follows Frank Lloyd Wright design styles. Jazmyne, who was closely involved in the set design planning, describes it in these terms, "It is just an absolutely gorgeous piece of carpentry. The host sits at a dining room table, a triangular table, but it has a dining room table feel. She has the questions in front of her, but she looks down at them frequently to prompt ideas, but the conversation is more of a chat. She adds a lot of her own ideas during the interview, too."

Jazmyne confesses to being a fan of Frank Lloyd Wright. "We have one of his houses, the Myer-May house, here in Grand Rapids. We went over there with the college set designer, who does all of the theatrical productions here. His design for the set was inspired by the photos we took of the house."

Hunting for Topics and Speakers

Jazmyne tends to follow her instincts on topics of interest, often ranging far beyond the usual kinds of community issues to topics of national import. "Right now," she said, "the one issue that is really burning me up is the price of pharmaceutical products and the whole monopoly behind them—the way the HMO system encourages the customer to not really care how much a drug costs. I just ask the pharmacy to fill my prescription and I don't even look at how much the drug costs."

Jazmyne relies on tried and true journalistic techniques to track down potential interview subjects for a topic that has gained her interest. "First, I do a little research at the index at the local paper, to see what has been written about a topic in the last year. Then, I try to find local people to talk about it, or maybe a professor at Calvin or Grand Valley who knows a lot about the topic. I might ask them to be on the show. I also might look for somebody who is coming through town who knows something about the subject—somebody who is either coming to speak at another location or somebody who is coming to Calvin."

Compelling People Not everyone works out ideally as an interview subject. "We had a consultant come this year," Jazmyne said, "who told us we need to work on getting people who are not just knowledgeable, but who are also compelling people. Just because somebody knows what they are talking about doesn't necessarily mean that they are good to watch on TV. My trick is to interview them over the phone and tell them that I am doing research for the topic. I don't care what they look like, because I don't want to buy into the whole media thing that everybody has to look perfect. But, if I find them interesting to talk to on the phone, then I know they will be interesting to talk to—usually—on TV. Although some people get pretty nervous on TV."

Calming Guests The art of relaxing guests is a primary requirement of getting a good interview. "Most people are pretty nervous when they first arrive for an interview," Jazmyne said. "We have them sit with the host for a good ten minutes on the set, chatting, talking about something other than the main topic until they loosen up."

Once the videotape starts rolling, the guests are generally more relaxed. Experienced hosts, such as Shirley Hoogstra, know how to gently ease the interviewee into natural conversation. Jazmyne sees this as a unique skill that is a bit different from simply being able to conduct an interesting interview.

Prompting Hosts Jazmyne uses the teleprompter if it is necessary to communicate with the host while the videotaping is underway. "We typically only use the teleprompter for the introduction at the very beginning when the guests are announced. Sometimes you can't even remember your own name in that first few minutes. So, we use that just for the beginning and the ending. But, if I want to say something to someone on stage, I type a note to them on the teleprompter, very brief, and then we have the camera operator wave at their teleprompter. The host looks over at it when the camera is not on her."

Directing Inner Compass

Rob Prince directs episodes of Inner Compass, as well as a number of other specialized productions for the college, using a crew largely composed of students. The equipment being used on the set has made a steady transition from analog cameras and switchers to digital tools, although Rob still professes a fondness for the softer image quality of some of the earlier analog equipment. Video editing takes place in the digital realm on a Macintosh running both Avid and Final Cut Pro.

Creating the Segments

Responsibilities for creating the video segments are shared between Jazmyne Fuentes and Rob Prince. "Rob and I split up the work," Jazmyne said. "I concentrate mostly on the content, lining up the topics and the guests, and scheduling the shoot. Rob hires the students who work for him as lab aids in the video editing lab. The lab helps teach students how to use the equipment and, as part of their job description, they are also there for *Inner Compass* shoots."

A crew of four students, designated through a weekly sign-up sheet according to availability, handle their share of the video production work, operating three cameras. Rob serves as the switcher while videotaping is taking place. Segments are taped in the new DeVos Communication Center, a well-equipped facility that can accommodate two individual sets and a chroma key wall. The first shooting of Inner Compass in this facility focused on Quentin Schultze discussing his recent book, *Habits of the High-Tech Heart*, an assessment of the ways in which electronic communication and accelerating technology change people and communities. A portion of this discussion appears on the DVD bundled with this book. The generic set shown in that interview has been replaced with a new *Inner Compass* set modeled after a nearby building that was designed by Frank Lloyd Wright.

From Analog to Digital

The transition to working in the new DeVos Communication Center has been accompanied by changes in the production equipment. "Our situation has varied a bit in the last semester since we've been shooting on two different sets," Rob said. "One was a completely digital DVCAM setup using DVCAM cameras running through digital switchers to a DVCAM deck. The other was an analog system using old Sony studio cameras running through an analog switcher into a DVCPRO deck."

Rob immediately noticed a difference in the image properties of the two systems, part of which, he thinks, might be a factor of lighting variations on the sets. "The digital cameras were so crisp and bright that I almost preferred our analog system. The analog system seemed to have a softer tone compared to the harshness of the newer cameras, but part of it could be the lighting on the set, too. But, surprisingly, we were quite happy in general with how the analog cameras looked compared to the digital."

Digital video editing has been used for several years within this production group. "The video," Rob said, "would be routed through analog cables into our DVCPRO deck and then from the DVCPRO unit directly into our Avid editor. From there, it would stay digital for the rest of production."

Rob found out the hard way about the disc incompatibilities between DVD+R and DVD-R. "We currently have a Panasonic DVD+R burner and Randy's duplication system doesn't like the DVD+R discs (it wants DVD-R). I bought the burner because it looked like a good deal and it had all the connectors that I needed."

The learning curve with any new equipment can be a challenge and digital video is no exception. "It has been nice how the DVCPRO deck can also work with the DVCAM tapes. You can switch it over through a menu so that you can digitize directly from DVCAM. That was something we originally didn't know we could do. You just have to learn as you go. I had a little experience with HD video when it first was coming out. I was working at a company that was just getting started with it—it was a nightmare. We couldn't even make dubs because the engineers would hook things up wrong or have the settings slightly off. It was a mess."

Rob has been directing productions at Calvin for three years and previously worked at the local PBS station, WGVU, which is the station that airs *Inner Compass*. Before that, he worked in Chicago for Kurtis Productions, a company that creates shows for A&E. Outside of his daily project responsibilities, Rob has been working with a partner on a documentary based on certain events in World War II, a project which he hopes to complete and distribute sometime in the next year or two.

Hand-Off for DVD Authoring

Randy Nieuwsma works closely with Jazmyne and Rob, generating a DVD from the DVCPRO tape that he receives following video editing. The content on tape can be directly imported into a DVD recorder, the Pioneer DVR-7000, where it then serves as the master for generating multiple copies.

"We need a broadcast copy," Jazmyne said. "We need a production copy, we need a copy for the library, and one for the host. We need a bunch done at once, primarily on DVD. In the past, Randy has been making VHS copies, but this last year, because of the transition, we started using DVCPRO as our master and then we would have VHS copies made for everybody. Now that we have this new system, we intended to go to DVD, but we got backlogged. So, Randy is spending the summer making all these copies. Next year, we will have them done on the spot. Randy is organized, but everybody doesn't supply him what he needs at the right time. Part of the problem is that we sometimes give him the DVCPRO tape and tell him that it has got to get in the mail to the PBS station in 35 minutes and it is a 26-minute tape. This usually means that we don't get all the duplicates made then and there. So, it gets put on the list of things to do."

Distributing Inner Compass to PBS on DVD

DVD has become the medium of choice for delivering shows to the local PBS affiliate, WGVU. The station, which relies on earlier broadcast equipment, receives the DVDs and then bumps them up to 1-inch videotape, their standard broadcast format.The broadcast engineer at WGVU notes that they don't have to do any special processing to use the DVD content—they basically set the audio and video recording levels from the color bars and audio tone included on the DVD-V disc. The engineer also notes that the quality differences between the DVDs and programs that he receives on 1-inch videotape are slight, but not objectionable. Quality differences may be associated with the fixed bitrate that is a factor that can't be adjusted on the real-time Panasonic DVD recorder that Calvin College uses for much of its work. Using a higher bitrate for the MPEG-2 video content would clearly result in significantly higher quality in the productions.

"DVD," Rob said, "is the one format that we have in common with WGVU. For a long time, the only format that we had in common was 3/4-inch videotape. They were working primarily from 1-inch. The 3/4-inch was the highest quality format that we had in common. They wanted us to buy a 1-inch player from them and we thought about buying them a DVCPRO deck or something similar. But, eventually, we all came around to DVD and that has been outstanding for us. We've improved our quality, and we are also spending one tenth the cost of tape material. Because 3/4-inch tapes are now basically antiques, they are getting very expensive. No one is using them anymore. We lucked out with the DVD."

Solving Archiving and Distribution Problems

As demonstrated by this case study, DVD can serve a number of roles for both video production, media distribution, and archiving in any type of media center or production facility. The long life of the media, the compact storage format, the ease of distribution, and the inexpensive blank media all add to the practical benefits of DVDs.

17

DVD Fundamentals

As much as you might like to completely ignore the technical details of DVDs, even the simplest authoring tasks require some knowledge of the file and directory structures, as well as awareness of the available DVD formats. You'll find that the more you get involved in making DVDs, the more useful this knowledge becomes in gaining the results you want. Since this book is designed to serve as a practical guide more than a set of specifications, if you're inclined to delve deeper into the mysteries of DVD, you might want to investigate other references that focus more on the technical aspects of this storage format. One useful place to start is *DVD Demystified* by Jim Taylor, the second edition published in 2001 by McGraw-Hill. Another valuable resource, available online, is the DVD Forum (*www.dvdforum.org*). Outside of exacting engineering applications, these two resources should satisfy most requirements for in-depth DVD technical knowledge.

The latest generation authoring software does a respectable job of shielding the user from many of the intricacies of file formats, encoding, menu structuring, and directory organization. However, if you're well grounded in this kind of information, it is easier to make informed decisions when planning a DVD project. Each of the DVD formats defined in five distinct books—Book A through E—has particular uses. Each format also suits some playback devices better than others or has limitations in terms of playback. Know these limitations and you'll be more capable of creating titles that can be successfully accessed by the widest audience.

This chapter provides a fundamental overview of the DVD layouts and formats and offers technical descriptions of other aspects of DVD technology that should prove useful to developers.

DVD Technology

In the early 1990's, the DVD Consortium (which is now the DVD Forum) set out to increase the storage capabilities of optical discs without altering the physical diameter that had become standardized with the audio compact disc. There were two obvious ways to do this:

- Compress more data within the span of an individual track
- Add additional layers within the disc itself

Since the work of the consortium has been codified as a set of standards, manufacturers have produced a wide array of DVD implementations. The implementations range from single-sided, single-pressed discs to multiple layered DVD-18 discs that store a substantial 17.9 Gigabytes of data.

Commercial DVD-Video discs, which deliver feature films to a prospective audience of 66-million DVD players in the U.S. as of mid-2003 (according to the DVD Entertainment Group), store as much as 17-billion bytes of data on the disc surface. This equates to a storage capability sufficient to handle nine hours of audio and video content (with appropriate compression on a multilayer disc) or 26 times more data than can be stored on an audio CD. The quality of the content surpasses other forms of video storage within the consumer marketplace, providing significantly better quality than the previous standard bearer—the Laserdisc.

The capability of DVDs to handle multichannel audio has led to the creation of *home theater* entertainment systems that position speakers around the viewer and produce a sensation of being in the midst of the onscreen action. Up to six-channel digital sound can be delivered, generally subdivided in this manner:

- Center channel for dialog
- Left channel for music
- Right channel for music
- Left rear channel for effects
- Right rear channel for effects
- Bass channel

Most of the information stored on a DVD-Video disc is in compressed format. MPEG-2 is typically used for the video content and Dolby Digital (AC3) for the audio content. The MPEG-2 digital audio format is also sometimes used for audio material stored on disc. These compression technologies make it possible to accommodate the prodigious storage capabilities of the media while delivering very high quality audio and video content.

DVD technology was designed from its inception to be backwards compatible with CDs. This presented an immediate technical challenge since the pits embedded in a CD occur at a different level on the disc surface. There are also as many as four possible discrete surfaces embedded within a DVD disc. The read laser in a DVD player must be able to adjust its focus to retrieve data from the various layers.

This technological difficulty was solved through the use of different lenses, including holographic lenses that can simultaneously focus on more than one distance at a time. The built-in backwards compatibility also embraces Video CD 2.0, an earlier format for distributing video material in MPEG-1 format. Video CDs achieved some success in Japan and Europe, but never made significant inroads in the United States.

Increasing Data Capacity

The first step in increasing capacity beyond the audio CD and CD-ROM is to tighten the spiral and reduce the size of the pits used to form the data impressions.

The second technique for increasing the storage capacity is to add additional layers—up to 4 layers total, two on each side of the disc. The laser beam focus is adjusted as necessary to access data on individual levels. Improved optics are required to achieve accurate data reading, as well as a reduction in the wavelength of the laser beam reflected off the disc surface.

The single-sided, single-layer DVD, as shown in Figure 17 - 1, consists of two 0.6 millimeter polycarbonate substrates that are bonded together to form a disc that is 1.2 millimeters thick. The stamped surface of a single-layer disc is coated with a layer of aluminum through a process called *sputtering*. This aluminum forms the reflective surface upon which the laser beam is used to detect the data pattern. This type of disc has a capacity of 4.7 Gigabytes and is referred to as a DVD-5 disc.

Two methods are typically used to bond the two substrates together:

- Hot-melt method: A thin coat of melted adhesive is spread over each substrate and then the two surfaces are bonded by means of a hydraulic ram. This is the least expensive method of bonding.
- UV method: A thin layer of lacquer is distributed over the disc surface to be bonded, either by rotating the disc or through silk-screening. Ultraviolet light is then applied to harden the lacquer. While this method is more expensive, it forms a bond that is resistant to temperature extremes.

The label surface of this disc can be printed using conventional printing techniques or silk-screening. Impressions molded into the blank substrate can also be used as a substitute for printing a label.

Figure 17 - 1 **Layers within a DVD-5 disc**

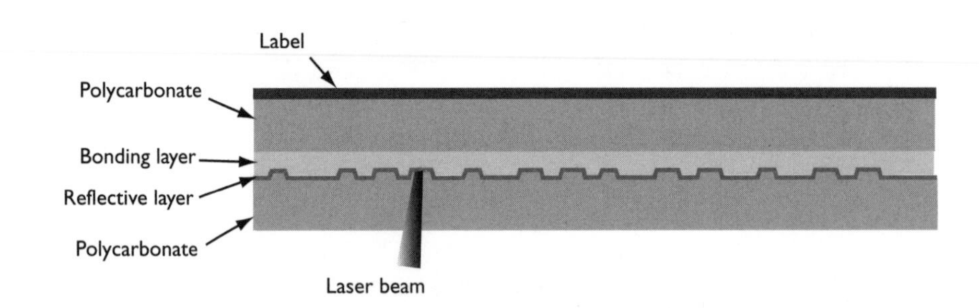

A DVD-9 utilizes two separate layers of data. The data layer closest to the laser is composed of a semi-reflective coating that enables the laser to focus on it for data reading or to focus through it to the next higher layer to read the data from the more reflective surface. The capacity of the DVD-9 disc, shown in Figure 17 - 2, is 8.5 Gigabytes.

The two substrates of a dual-layer disc are bonded together using one of two methods:

- By means of an optically transparent adhesive film that affixes the layers to each other

- Through a photopolymer material that combines the second layer on top of the first on a single substrate, which is then bonded to the blank substrate

Conventional methods of lithographic printing or screened printing can be used to complete the label.

Figure 17 - 2 **Layers within a DVD-9 disc**

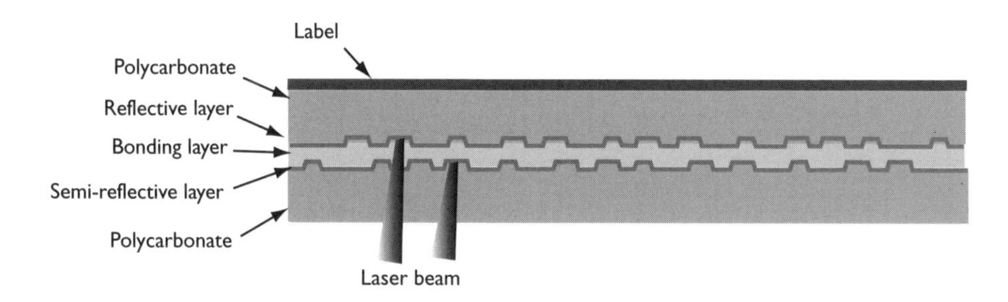

A DVD-10 disc has data on both substrates, using reflective layers on each so that the disc must be physically flipped over in the DVD player or DVD-ROM drive in order to read the data on the second side. This format has a capacity of 9.4 Gigabytes. Since the laser must be directed through both surfaces of the disc, no label is applied to either surface—this is a quick way to recognize this particular format, shown in Figure 17 - 3.

Figure 17 - 3 **Layers within a DVD-10 disc**

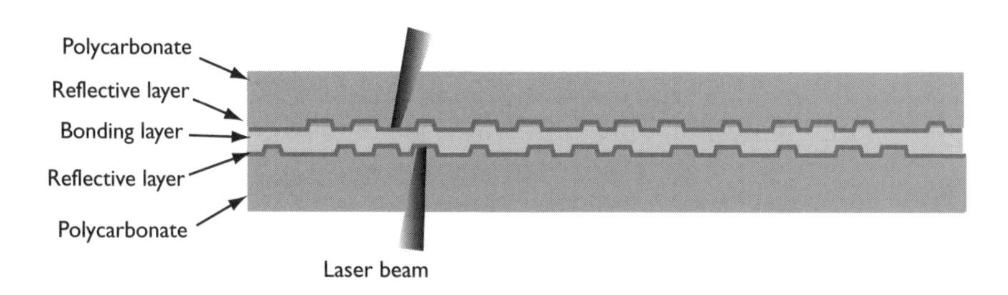

The most complex DVD format, and the most expensive to manufacture, is the DVD-18 disc, which includes both dual sides and dual layers, as shown in Figure 17 - 4. The two layers on each side must be manufactured on a single substrate. One layer is created on a substrate using a conventional stamper to produce the data pattern and then a second stamper creates a data image on a photopolymer material, which is then affixed to the substrate. This same process is followed for the second substrate, which also contains two layers of data, providing a total capacity of 17.9 Gigabytes. As with a DVD-10 disc, this type of disc must be turned over in the player or the DVD-ROM drive in order to access the data on the second side. Similarly, neither surface can be printed with a label, since the surfaces must offer clear access to the laser for data reading.

Figure 17 - 4 **Layers within a DVD-18 disc**

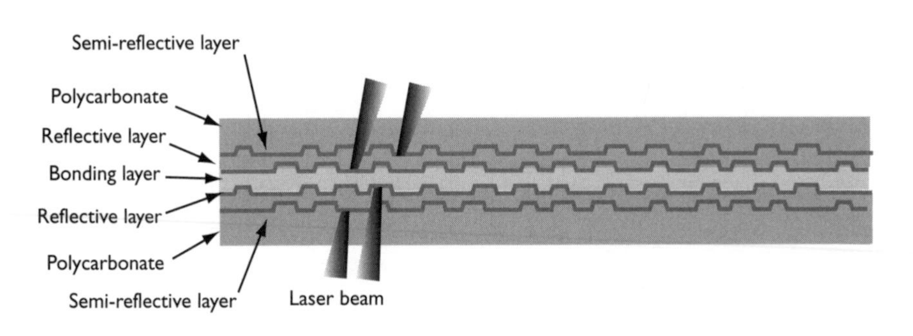

Storing Content on DVD

The addition of new layers to a disc, each with its own microscopic spiral of data, adds to the complexity of manufacturing. For this reason, and the additional costs associated with multilayer disc manufacturing, many commercially released DVD titles utilize only a single layer. This is sufficient to store a typical 135-minute film at a degree of compression that provides a reasonably high quality image. For longer works, or for including multiple movie titles on a single disc, additional layers can be added to the disc to achieve the desired capacity. Approximately 2 Gigabytes of storage is required for each hour of video material compressed with MPEG-2. Title producers can determine the amount of storage needed for a project and then master the disc accordingly.

Unlike a compact disc, which employs a single substrate, a DVD is composed of two 0.6 millimeter substrates that are combined to increase the rigidity. The additional rigidity is also important for the overall disc balance and for reducing the amount of wobble while the disc is spinning. These are both critical characteristics for ensuring accurate reading of data from the medium.

Reading All Types of Discs

In many ways, the DVD is an extension of the data storage techniques originally perfected for CD-ROMs and CDs, but with its own special characteristics. Just as the vast majority of CD-ROM drives are capable of playing audio CDs, DVD-ROM drives have been designed to be backward compatible with CDs and CD-ROMs. The following table summarizes the similarities and differences between the two types of discs.

Table 2: Comparison of CDs and DVDs

	CD	DVD
Diameter	120 millimeters	120 millimeters
Thickness	1.2 millimeters	1.2 millimeters
Data capacity	680 Megabytes	4700 Megabytes
Layers	1	1, 2, 4
Track pitch	1.6 nanometers	0.74 nanometers
Minimum pit length	0.834 nanometers	0.40 nanometers
Laser wavelength	780 nanometers	640 nanometers

The optical pickup designed for use in a DVD unit is mounted on an arm that positions the laser beneath the disc surface during playback. As you can see from the previous table, the required laser wavelength is different for CDs than it is for DVDs. One technique for handling this difference is to use a twin-laser pickup that features completely separate laser and lens fixtures. If the DVD player or DVD-ROM drive is attempting to read a CD or CD-R, it uses the fixture optimized with a laser wavelength for these media types. For DVDs, DVD-Rs, or DVD-ROMs, the unit employs the laser and lens with the wavelength optimized for DVD media.

A focusing control adjusts the depth of focus to be able to read the individual DVD layers. For a DVD, layer 0 is about 0.55 millimeters above the bottom surface of the disc. The second layer, if present, is another 55 micrometers higher. In comparison, data that is embedded in a CD or CD-ROM appears approximately 1.15 millimeters above the bottom surface of the disc. This difference, as well as the different laser wavelengths required for reading the data, is the reason that separate lens and laser fixtures must be used for DVD and CD media.

Recordable Forms of DVD

Recordable forms of DVD come in several varieties with varying degrees of compatibility with existing playback equipment. The primary categories are:

- DVD-R—as defined in Book D of the DVD Forum's specifications, this is the write-once form of the media with a standard capacity of 4.7 Gigabytes. The recording surface is a dye layer. Recorded discs are playable in many standard DVD-ROM drives. DVD-R drives can be designed for Authoring or General Use DVD-R media. Authoring media is preferred when masters are being created for replication.
- DVD+R—developed by Sony and Philips outside the purview of the DVD Forum, this format was designed to facilitate wider compatibility of recordable discs with DVD players. The specification for this format is maintained by the DVD+RW Alliance.
- DVD-RW—similar to CD-RW, this approach relies on phase-change technology to support the erasing and rewriting of data. The standard capacity is 4.7 Gigabytes.
- DVD-RAM—defined in Book E of the DVD Forum, this format also uses a phase-change recording layer, which may be single-sided or double-sided. Capacity is 2.6 Gigabytes per side. The RAM stands for Random Access Memory.
- DVD+RW—also developed by the DVD+RW Alliance, this format uses lossless linking technology, a technique that allows the recorder to accurately stop and start data write operations. The media for DVD+RW handles up to 4.7 Gigabytes and offers approximately 1,000 rewrites.

Early in the design process for the writable form of CD-ROM (CD-RW), progress was hampered by the incompatibility of the rewritable discs with the majority of CD-ROM drives. This problem was overcome by the introduction of the Multiread specification for CD-ROM drives, an extension which ensured that drives certified as MultiRead ready could retrieve data from CD-RW discs.

A similar effort has created a Super MultiRead specification that encompasses the range of recordable forms of DVD, so that the various formats will be readable in Super MultiRead-certified DVD-ROM drives and players. In general, the latest generation of players and DVD-ROM drives handle the full range of stamped, duplicated, and rewritable media quite well. Older players and drives, however, are unpredictable and much more likely to reject the new recordable formats.

DVD-R for Write-Once Applications

DVD-R media and recorders can produce discs that are suitable for premastering of DVD-ROMs or DVD-Videos, as well as discs intended for data distribution and exchange, document imaging, and archiving. Initially, many replication services required DVD masters to be submitted on the Authoring version of the media, which was the original type of recordable disc designed for professional use. The General Use media, designed for consumer use, is now accepted by some replicators, but, as a whole, the industry still favors Digital Linear Tape (DLT) for submissions. General Use DVD-R hardware and media do not allow data to be written to the lead-in area on the disc, which prevents the use of CSS copy protection. However, another form of copy protection, CPRM, can be used with General Use media.

The specification crafted by Working Group 6 (WG-6) of the DVD Forum includes provisions for single-layer, single-sided media or single-layer, dual-sided media. As with CD-R media, both 12-centimeter and 8-centimeter disc sizes are supported, although the large majority of applications rely on the more common 12-centimeter disc size.

Two polycarbonate substrates—one containing a dye layer and reflective coating and the other blank—are bonded together to produce a 1.2-millimeter thick disc for single-sided DVD-R applications. The first forms of DVD-R media used only cyanine dye, which appears violet on the recording side of the recordable disc.

A spiral pregroove extends from the center of the disc to the outer diameter to act as a guide for the laser during recording. A slight wobble in the pregroove in a pre-established pattern generates a frequency used as a carrier signal; the timing information helps regulate servo motors, tracking of the laser assembly, and focus of the beam. Land pre-pits molded into the substrate provide address information and pre-recorded data, used to initiate write operations.

Pulsed laser beams directed at the dye in the pregroove form impressions by searing variable length marks in the dye surface. These marks, consisting of deformation of the substrate material and bleaching of the dye, serve the same purpose as pits in a pressed DVD disc. Areas in the pre-groove that are not exposed to the pulsed laser are interpreted as lands.

DVD-R recording requires a more complex write strategy to establish the appropriate lengths for the pits, which are approximately half the size of those on a CD-R disc. The spacing between the pits and lands within the spiral data pattern is also significantly less than on a CD-R disc. To compensate for the extra precision required during write operations, the laser pulses are very carefully controlled, both in terms of intensity and duration. During recording, the laser is rapidly modulated between the power setting required for writing and the setting used for reading to avoid overheating the media surface and to regulate the size of the mark seared in the dye. A technique known as Optimum Power Calibration (OPC) is used to perform test write operations to a specified calibration area on the recordable media surface and then to read back the test pattern and adjust the laser power settings to match the recorder to the media. Given the extra precision required for recordable operations using DVD-R, this feature becomes a highly desirable addition to any recorder and helps ensure the most consistent results when performing disc recording.

First generation DVD-R media offered capacities of 3.95 Gigabytes and approximately 3.68 Gigabytes of usable space (considering the overhead required for lead-in and lead-out areas and other file system data). The recordable capacity of the second generation DVD-R discs is 4.7 Gigabytes, of which approximately 4.38 Gigabytes is available for data storage.

Data transfer rates for recording DVD discs were initially based on a nominal 1.32 Megabytes per second rate, which is considered 1x speed. At this data transfer rate, completing the recording of a 4.7GB DVD-R disc requires slightly less than an hour. Current generation DVD-R equipment can accelerate the write process using 4x speeds.

DVD-R serves an important role in project prototyping for developers and title producers, since it is designed by definition to be playable in standard DVD players and DVD-ROM drives. Early recorder costs were in the $17,000 range, but as was the case with CD-R equipment, costs have been declining dramatically. Second generation equipment, such as the Pioneer DVD-S201, dropped to the $5000 mark. Authoring caliber DVD-R drives can now be purchased for about half that amount. General Use DVD-R drives can be found for as little as $250. Blank media costs have dropped from approximately $40 for 4.7GB discs and $35 for 3.95GB discs to about $5 for Authoring media and $1 for General Use media.

Table 3: Differences in Recordable DVD-R Types

Property	Version 1.0	Authoring use	General use
Number of sides	1 or 2	1 or 2	1
Data capacity	3.95GB	4.7GB per side	4.7GB
Recording method	Organic dye layer		
Laser wavelength	635/650nm	635nm	650nm
Minimum pit length	0.44 microns	0.40 microns	0.40 microns
Track pitch	0.80 microns	0.74 microns	0.74 microns
Serialization for CPRM		No	Yes
Pre-recording		No	Yes
Track format	Wobble pre-groove		
Modulation	8/16 modulation		
Error correction	Reed-Solomon Product Code		

Those who followed the development of CD-R technology witnessed the difficulty inherent in maintaining compatibility given the many variables in recordable media, playback equipment, recorders, premastering software, and so on. It took several years for all these varying characteristics to be tamed and controlled in such a way that recorded discs could be freely distributed among the vast majority of CD-ROM drives. A similar evolution is taking place with DVD-R as manufacturers, engineers, and developers refine the tools and techniques used to burn data in discs. Early adopters of this technology faced a variety of trials and tribulations as the compatibility problems were worked out. Today, manufacturers and standards organizations have solved many of the compatibility issues. Users of both DVD-R and DVD+R media can be expected to see compatibility somewhere around 95 percent with DVD players currently being manufactured. Compatibility with earlier players, however, is still a hit and miss affair.

DVD Formats: The Five Books

The Digital Versatile Disc—DVD—suits its name very well. Since the DVD-ROM and DVD-Video specifications were introduced as version 1.0 in 1996, DVDs have become a dominant medium for information storage, both in the entertainment world as a vehicle for distributing feature films and other movies, as well as in the computer world where recordable DVD drives have become standard equipment on many systems. The versatility of this storage medium can be demonstrated by looking at the range of uses, which has led to DVDs exceeding the growth patterns of any previous electronics technology.

Since the DVD-ROM and DVD-Video formats were first introduced in 1996, a variety of extended formats have emerged, including several recordable and rewritable types and a DVD-Audio format. Even as the installed base of players continues to grow rapidly, industry leaders are working on a new format to support high-definition video content.

The order in which these standards have been developed follows the sequence of the Books, A through E.

An Evolving Set of Standards

If the DVD standards had been developed logically and methodically by a non-partial committee of neutral participants, we'd probably have a more consistent framework for the growth of DVD. Instead, the standards evolution has been pushed and pulled by groups of electronics manufacturers and entertainment conglomerates, each trying to wield their influence on the process for the gain of their organization. DVD has been pulled and tugged in every conceivable direction by forces representing sometimes diametrically opposed viewpoints.

At this point, the situation for the primary standards is quite stable, although much activity is still going on trying to produce DVDs with audio that can be played in CD players, double-sided hybrid DVD-Audio discs, and other kinds of variations, such as high-definition DVDs. Today, most new DVD-ROM drives will currently read commercially pressed CD-ROMs, as well as CD-R, and many will handle CD-RW as well. DVD players have reached a production volume where the prices have dropped below the $100 mark for entry level units and most of the incompatibilities exhibited by early releases of DVD-Video titles have been eliminated. Consumers, initially wary of the medium from the constant flow of negative information from the press, are purchasing players and DVD-ROM drives in record numbers. The once sluggish growth curve is now showing record growth patterns. By the middle of 2003, more than 66-million DVD players have been sold in the U.S. and almost 1.8-billion discs have been shipped in North America since the first DVD was sold.

DVD-ROM

The DVD-ROM is the computer data version of the digital versatile disc. As with the transition of the audio CD to CD-ROM, the DVD-ROM includes extended support for error detection and correction to allow it to be successfully applied to computer applications where a missing bit can freeze an application. Unlike the audio CD, however, DVD-ROM is considered the starting point of a succession of standards that includes:

- Book A: DVD-ROM
- Book B: DVD-Video
- Book C: DVD-Audio
- Book D: DVD-R (write-once)
- Book E: DVD-RAM (rewritable)

All of these formats were devised with the goal to create an optical disc format that supported a significantly higher storage capacity than the CD-ROM. From the beginning, the standards bodies involved in developing and refining DVD specifications wanted individual formats to support audio, video, and computer uses, as well as a writable storage format. The five books developed around these objectives.

Data Storage Techniques

DVD-ROM stores data in user sectors, each consisting of 2064 bytes that are organized to support an error correction scheme. Of this total, 16 bytes are reserved for address information, error correction, and copy protection, leaving 2048 bytes for data. Data sectors are structured as 12 individual rows each consisting of 172 bytes. The beginning of each data sector contains 16 bytes of data, subdivided as follows:

- 4 bytes of identification data representing the sector ID
- 2 bytes of ID error detection data
- 6 bytes of copy protection data

The data sector is concluded with an additional code:

- 4 bytes consisting of an error detection code

The organization of the bytes within a DVD-ROM data sector is shown in Figure 17 - 5.

Figure 17 - 5 **Organization of a DVD-ROM data sector**

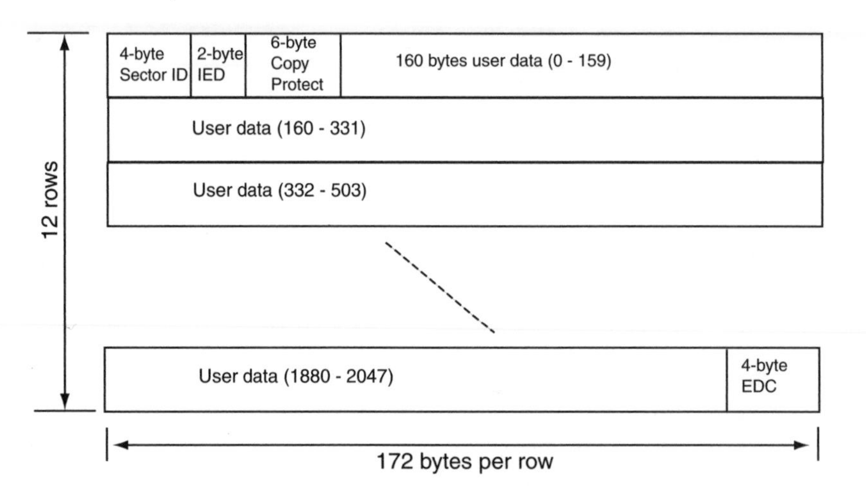

Rows of 16 data sectors are interleaved together and structured as blocks for applying error correction codes. A 16-byte Reed-Solomon code is generated for each of the 172 columns within the block. A 10-byte inner-parity Reed-Solomon code is generated for each of the 208 rows of the block. These codes are appended to the data, where they provide a flexible and robust means for detecting read errors within the data.

Through processing by means of 8/16 modulation, each data bytes is doubled to 16 bits, which produces a physical sector size of 4836 bytes. These bytes are generated on the disc surface, row-by-row, as channel data. As with data embedded on a CD-ROM, the Non-Return to Zero Inverted (NRZI) encoded method is used: transitions detected by the laser from a pit to a land are interpreted as binary ones; the absences of transitions are interpreted as binary zeros.

Channel data from the DVD media is transferred at the rate of 26.16M bps, which is then reduced by half by the application of the 16/8 demodulation process, resulting in a rate of 13.08M bps. After the adjusted overhead of error correction, the data transfer rate is a steady 11.08M bps. From the perspective of a DVD-ROM drive, data is transferred in logical units, each unit consisting of 2048 bytes.

UDF

One of the significant additions to the DVD standard is the widespread adoption of the Universal Data Format (UDF) as the means for dealing with files and volumes stored on disc. Although, theoretically, the data regions of a DVD-ROM can contain any type of data, most companies and organizations have followed the lead initiated by OSTA, the Optical Disc Storage Association, and adopted UDF for mapping file and volume structures.

UDF refines a more broad framework constructed by the International Standards Organization in ISO 13346. UDF places limitations on ISO 13346, defining a structure that supports optional multivolume and multipartition divisions on a disc. This allows DVD-ROMs that include filename translations between platforms and support for extended attributes, such as the resource forks, icons, and file/creator types that are familiar to Macintosh users.

Within the UDF standard, the following platforms are supported:

- DOS
- OS/2
- MacOS and MacOS X
- Windows 98/NT/2000/XP
- UNIX

The ability to partition discs for different playback equipment makes it possible for manufacturers and title producers to provide content that is specific to a playback platform. In the same manner that Enhanced CDs include both audio content for a standard CD player and computer data for playback in a CD-ROM drive, a DVD disc can have player and computer partitions. The digital video content and a wide range of interactive content can be included in the partition designed for the DVD player; this can include a director's commentary on a film's production issues, alternative language editions, and so on. The DVD-ROM partition can include items such as:

- Interactive multimedia content
- Games
- Screen savers
- Links to Web sites where related information about the title is available
- Background information on the cast and crew or design team
- Many other similar kinds of content

Most computer platforms also include some form of video player software for DVD, allowing films to be played on the desktop with the same crisp resolution and fluid playback that you will find on a dedicated player. The key element to making this happen is an MPEG decoder in the playback system, either implemented in hardware or as a standalone software component. Most DVD-ROM drives that are installed as original equipment in new PCs include MPEG decoding hardware as part of the package. If your DVD-ROM drive lacks this hardware, you must obtain a software decoder or you will be unable to play back DVD-Video titles.

Decoders embedded in hardware relieve the playback system processor of the burden of performing the intensive conversion process on the fly, which clearly improves playback performance. Whenever possible, hardware decoding is the preferred approach. The other component necessary to play back DVD-Video content on a PC is that the hardware or software decoder must be able to handle the encryption scheme that is built into DVD-Video discs for copy protection.

DVD-Video

DVD-Video is a specialized form of DVD-ROM that is tailored to the presentation of very high quality audio and video content optimized for set-top players. This is the format that the film studios, video publishers, and consumer electronics manufacturers have been backing as the predominant delivery medium for motion pictures in the new millennium. From a hesitant beginning, the format has caught hold solidly. Given the ongoing backing of so many of the major corporations involved in entertainment and consumer electronics, continuing success is assured.

DVD-Video relies the compression capabilities of MPEG-2 to provide, minimally, 94 minutes of video playback, but up to several hours of playback using the higher capacity formats. MPEG-1 video content can be included on a DVD-Video, but this option is rarely used because of the reduced quality of the compressed video.

The DVD-Video format was devised to support playback on the full range of standard NTSC and PAL television displays using analog data connections, ensuring broad compatibility with the installed television sets around the world. Most current-generation DVD players also include additional digital data connections, including S-Video and optical connections for more advanced televisions that support this form of signal input. The high-definition televisions that are appearing in the market can often take advantage of this digital interface.

Multichannel digital audio support is also an inherent feature of this medium, allowing audio content to be played on standard stereo audio systems, as well as more elaborate home theater systems. Up to eight channels of Dolby Digital audio provide the potential for excellent spatial orientation and rich, full sound to accompany videos.

To achieve the best results for audio works, specialized mastering tools must be used to separate the audio tracks into individual components. If this is not done effectively, the resulting audio performance can exhibit annoying characteristics, such as drifting orientation of the dialog track or poor signal clarity.

File Formats under DVD-Video

A DVD-Video disc can contain data for both playback on a DVD player and additional data content designed for computer playback. Based on the UDF specification, a specific directory is designated to store the video files, VIDEO_TS. An informational file titled VIDEO_TS.IFO must also be present. VIDEO_TS.IFO stores the video manager title set, which contains the contents of the Main Menu that appears when the DVD is mounted in the player.

Other title set information is contained in additional .IFO files and backup copies containing this same information are also maintained. Up to 10 video object block (.VOB) files can be created for each title that appears on the disc; these become the logical divisions by which the disc content can be navigated. Directories and files not intended for use by the

DVD player must be stored after the DVD-V data; these files are typically ignored by the player.

The original UDF standard was modified with an appendix, the MicroUDF, to simplify the recommended requirements that must be met by a DVD player, in the interests of encouraging widespread manufacturing of consumer-level playback equipment.

Appendix 6.9 of the UDF standard includes the following provisions:

- No multisession formats or boot descriptors are permitted on a DVD disc.
- Individual files must be contiguous and smaller than 1 Gigabyte.
- No more than one logical volume, one partition, and one file set can be included on a single-side of a disc.
- DVD players should support UDF in anticipation of ISO 9660 being gradually phased out.
- No more than 8 bits per character should be allocated for file and directory names.
- Aliases can not be used for linking.

The basic file structure used on a DVD-V disc is shown in the following figure.

Figure 17 - 6 File Structure for DVD-V

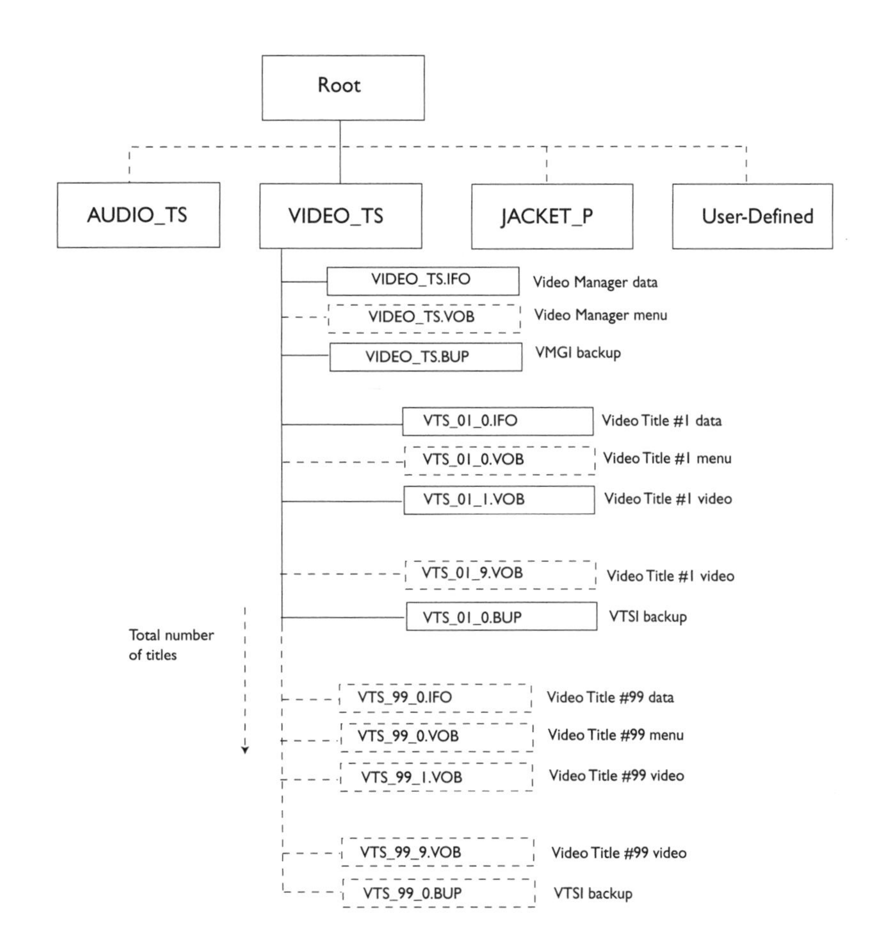

Navigating DVD-V Content

Unlike linear systems for video playback, such as film or videotape, DVD offers random access to the content on disc. Once is a disc is inserted in a player, a main menu can be accessed, which guides the user through the various types of content available on the disc. This may include a chapter view of a movie (allowing the viewer to jump to a particular point in a film), production information in video form, still images, audio content, interviews with directors and cast, and so on.

This type of content is handled by a presentation engine and a navigation engine contained in the player. This provides the equivalent to a computer user interface to the viewer, allowing someone to control the selection of the different elements for playback or to activate certain features available on the DVD. For example, a user might select an audio commentary to be played at the same time the video is being presented.

Figure 17 - 7 **Navigation and presentation engines**

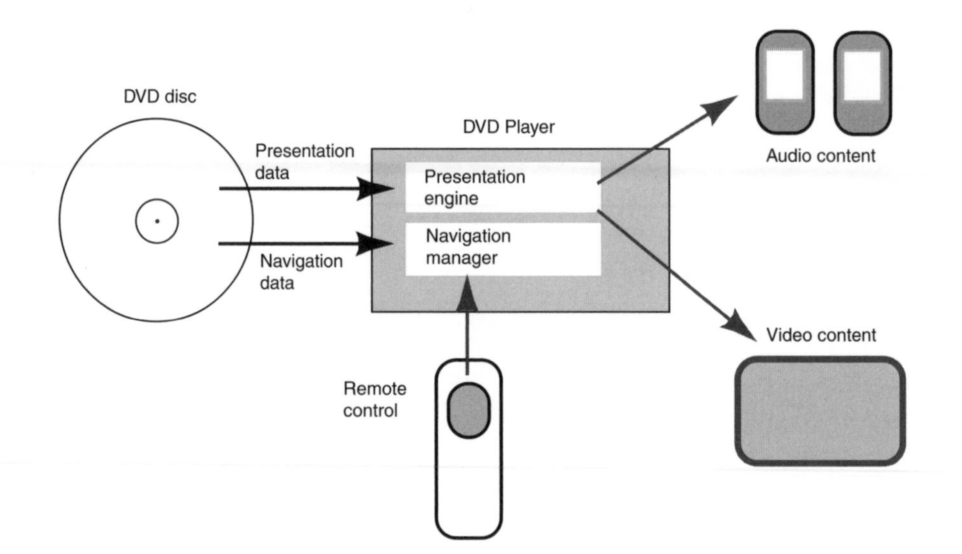

The actual content on the discs, from the perspective of the user interface, consists of:

- *Titles*, which may be films, videos, or album material
- *Parts of titles*, which may be chapters or individual songs

Each DVD can contain up to 99 discrete titles. Each title can be further subdivided into 10 chapters. For example, a disc containing the three-part television miniseries Stephen King's *Storm of the Century* might contain an individual title for each part and then 8 chapters within each title, allowing selection of any part of the series within 12 to 15 minute increments.

Optionally, a DVD can contain only a single title. For more complex projects, titles can also be nested within other titles and each title can have its own title menu. Viewers navigate between the various options using a remote control that features directional arrow keys and a select button, as well as the usual assortment of controls that one might find on a VCR remote, such as pause, scan, forward, back, and so on.

A presentation can include many different optional features that are handled by the navigation manager. For example, a project might incorporate a second camera angle that could be enabled through the menu to allow viewing of the content from a different perspective. Subtitles can be included for display when viewing foreign films. If any of these kinds of features are not required, the DVD author simply bypasses them within the authoring application.

Authoring for DVD-V

Authoring tools available for creating DVD-V content range from simple consumer-oriented applications for creating DVDs of family events to expensive, multi-featured professional-caliber tools. As both hardware and software tailored for creating DVDs has become less expensive, the trend has been towards an overall reduction in the cost of tools. Very sophisticated DVD authoring applications, such as Apple DVD Studio Pro v2.0 and Sony Vegas Video+DVD, can be purchased for less than $1000.

Encoders for producing MPEG-2 video range in price from software-only solutions suitable for both home and professional applications to elaborate standalone hardware systems with high-speed processors. Hardware approaches can deliver consistent, high quality video at rates that support industrial and commercial applications. More exacting video compression, however, is often performed using software encoders, which can be used to encode difficult variations in video frames by adjusting parameters and extending compression time. Many authoring packages include a built-in MPEG-2 encoder that can prepare raw video content prior to burning a DVD-R or DVD+R disc.

DVD-Audio

DVD-Audio is being positioned as a replacement for CD-DA (Compact Disc - Digital Audio), though the launch of players and titles has been slowed by a number of issues and acceptance by consumers has been much more gradual than for DVD-Video. DVD-Audio provides a minimum of 74 minutes of high-resolution Surround Sound. This format also supports a number of extended features, such as video content and simple interactivity. DVD-Audio discs support 5.1 channel Dolby Digital audio content, enabling fully equipped home entertainment systems with multiple speakers designed for enjoying DVD-Video sound to play back DVD-Audio discs at an equivalent level of quality.

As it stands, the DVD-Audio standard provides several significant enhancements over audio delivered via CD:

- Multichannel audio: the availability of 5.1 channels supports surround sound, encoded as Dolby Digital or MPEG-2 audio data
- Higher sampling rates: digital audio content can be sampled at rates up to 96K bps, producing greater frequency response
- Larger sample sizes: dedicating more bits to the sample sizes provides extended dynamic range and increased depth-of-field
- Display of album titles, lyrics, artist's names, and song titles during playback

The format for DVD-Audio includes the capability of including a DVD-Video sector. This approach makes it possible to create a high-capacity DVD equivalent to an Enhanced CD, with the audio material supplemented by video material and URLs that could be linked during playback in a DVD-ROM player. Another way of looking at it is that DVD-Audio combines the features of three separate DVD formats: combining audio, video, and computer data within a single framework. Most DVD-Audio players are also capable of playing DVD-Video content and most DVD-Video players can play back DVD-Audio discs. A number of major music companies, as well as some enterprising independents, have released titles on DVD-Audio discs and [as of late 2003] there are more than 1000 titles in the market.

The Working Group 4 (WG-4) of the DVD Forum continues to refine the DVD-Audio standard, but the fundamental elements of this standard have at reached the point where player manufacturers have felt confident enough to design and engineer the playback equipment. A fair amount of research and development is still taking place to try to design alternative formats, such as Super Audio CD (SACD), which includes layers that permit playback in CD players at CD quality. The high-quality audio (and optional surround sound content), however, require specially equipped DVD-Video players compatible with SACD.

Authoring of DVD-Audio content requires both a different set of tools and a different mindset for the development community. The abundant storage capabilities encourage titles that include not only audio, but video clips and computer data as well. However, as has been the case with Enhanced CDs, the extra costs associated with this type of development may limit the number of titles that include bonus content and extras. Different hardware and software tools are required to be able to record and mix the audio for 5.1 channel playback. The potential, however, for DVD-Audio is enormous. Audiophiles and music lovers looking for a way to experience more rich and vibrant sound, with the added benefit of audio and video content, should eventually be won over by the advantages of the format as more types of playback equipment and a greater selection of titles reach the marketplace.

DVD-R

If you want to burn a DVD disc designed for the widest possible playback, including current DVD players and DVD-ROM drives, DVD-R is the best choice. DVD-R is the write-once member of the writable family tree. The other rewritable formats currently lag behind with a number of compatibility issues that can create problems with some playback equipment. The DVD Forum maintains the DVD-R standard, while a competing organization, the DVD+RW Alliance has introduced DVD+R, which currently also has a very high degree of compatibility with more modern DVD players.

The DVD Forum has defined two individual categories in the DVD-R Book version 1.9 to accommodate different uses of DVD-R:

- DVD-R for Authoring: utilizes a laser wavelength for writing of 635 nanometers. This variation is designed for authoring only.
- DVD-R for General: utilizes a laser wavelength for writing of 650 nanometers, suitable for write operations intended for general uses.

Depending on the selection of recordable media, a DVD-R can handle up to 3.95GB or 4.7GB of data per side. Double-sided media boost the ultimate recording capacity to 9.4GB per disc.

The extra capacity of a DVD-R disc derives from the use of a red laser with a 635 or 650nm wavelength and an objective lens with a numerical aperture of 0.6. The smallest recorded mark possible on a CD-R disc is 0.834 micrometers, while DVD-R pits can be as small as 0.4 micrometers. This results in an increase of almost seven times the data density of the DVD-R disc over what can be stored on CD-R media.

The track pitch—the distance between two adjacent tracks on the spiral of data—is also much tighter for DVD-R than CD-R: 0.74 microns for the 4.7GB media and 0.8 microns for the 3.95GB media. All of these factors—the laser wavelength, pit size, track pitch—make it possible to achieve the high storage capacities available on DVD-R.

Playback Compatibility

DVD-R drives offer the advantage of broad playback compatibility, which includes most of the playback equipment available for DVD format discs. Discs can be recorded in either DVD-V or DVD-ROM format.

A properly recorded DVD-Video disc produced on a DVD-R machine can be played on:

- Most standalone DVD-Video players
- DVD-ROM drives installed in a host computer as long as they are equipped with either a hardware-based MPEG decoder or a software decoder that performs the same function

DVD-ROM discs that are recorded on a DVD-R drive can generally be played back on:

- A DVD-ROM drive installed in a host computer, without built-in MPEG decoding
- A DVD-ROM drive installed in a host computer with either hardware or software MPEG decoding (if access to DVD-Video material is required)

Both MPEG-1 and MPEG-2 decoders are available in both hardware and software implementations. The earlier MPEG-1 encoding method was first introduced for storing video content on CD-ROM. Commercial Video CDs, including movie titles, were introduced in the market shortly after multimedia CD-ROMs became popular. The Video CD format fared more successfully in Europe and Japan than in the U.S. Many DVD players can also play back Video CDs, if you happen to have any of the released titles in your collection. Most MPEG-2 decoders can handle both the MPEG-2 and MPEG-1 data formats.

File System for DVD-ROM

As described earlier in the chapter, the file system for DVD discs is much different than the fragmented system that was designed for CD-ROMs as they evolved and grew to encompass many different types of data formats. Instead, the more unified UDF structure was applied to all types of DVD discs and this file system is suitable for all forms of content and any type of file format designed for storage on optical disc. It is also designed to be adaptable to all the major computer operating systems.

To maintain backwards compatibility with earlier computers and operating systems that are not designed to read UDF, the UDF Bridge file system was designed. UDF Bridge is a hybrid system that includes support for discs recorded using the original ISO-9660 file system that originated with CD-ROMs. UDF Bridge also provides full support for discs containing files structured under UDF, as well.

UDF Bridge maintains an important optical disc convention that has been followed ever since the Yellow Book CD-ROM format was introduced as a means of extending the compact disc to include computer data, as well as audio data. Computer-based playback devices and decoders, as much as possible, have been designed to read all earlier formats. CD-ROM drives were designed to read audio CDs. DVD-ROM drives, from the beginning, were equipped to be able to manage CD-ROM playback. The UDF Bridge format makes it possible to provide backwards compatibility with ISO 9660-based readers while also offering the benefits of the UDF structure.

In comparison, playback devices targeted for the entertainment industry and consumer use, such as audio CD players and DVD playback equipment, usually only accommodate a single format. The Enhanced CD format was designed as a hybrid format—containing both audio and computer data. This allows CD players to play the audio present on the compact disc, while the computer data files could only be read if the disc were inserted in a PC CD-ROM drive. Similarly, DVD-Video discs often contain content designed for the PC, which is essentially ignored by the DVD playback equipment. The additional data only becomes available if you insert the disc in a DVD-ROM drive. A file folder labeled ROM is recognized as the storage receptacle for content designed for computer playback.

Writing to DVD-R Media

Anyone familiar with CD recording applications and equipment will recognize the very close parallels to DVD-R write operations. The recording process is handled by an application running on the host computer connected to the DVD-R unit. For example, Roxio Easy CD Creator 5 is such an application. The application lets the person doing the recording select the files and organize them for the write operation. The application then manages the actual recording process, controlling the DVD recorder until the write operation is complete.

Disc-at-Once

Like CD-R discs, DVD-R discs can be fully written in one complete operation—known as Disc-at-Once—or written incrementally over several individual sessions. Unlike CD-R discs, however, data written to DVD-R discs occurs in a slightly different sequence. The recording application first produces a lead-in area, followed by user data area, followed by a lead-out area. The lead-in and lead-out areas contain information that allows the DVD player or DVD-ROM drive to properly access the full range of data. These same two areas also appear on commercially pressed DVD disc—they are essential to playback. The user data, sandwiched in between these two regions, can vary from a 32Kb block—the minimum amount of data that can be recorded—up to the maximum capacity of the recordable media: 4.7GB.

In comparison, CD-R discs are written in a different sequence. The user data regions are recorded first. Next comes the lead-in area and the table of contents. The write operation is concluded typically by writing the lead-out area.

Write Operations

During write operations, the host computer must be able to deliver the data at 11.08 Megabits per second to prevent any interruptions in the data pattern being recorded to disc. Buffering is used in the DVD-R drive to compensate for any interruptions in the flow of data.

When incremental write operations are being performed, the complete file system data is not available until all of the individual writes have been finished, so the disc must be finalized before it can be read in any device other than another DVD-R drive. Finalization calculates and records the data contained in the lead-in and lead-out areas and then records this information to disc. Once a disc has been finalized, it can be read by other DVD playback devices. No further data can be recorded after a disc has been finalized.

At the recording speeds of the current generation of 4x equipment, a 4.7GB disc can be recorded in about 15 minutes.

Uses for DVD-R

The lowering costs of DVD-R recorders may persuade many corporate and individual users who are looking for archival storage or backup applications, even if they have no intention of producing discs for commercial replication. This category of users may also find a viable solution in the rewritable formats discussed in the next section.

The broad compatibility of DVD-R makes it a medium well suited for high-volume data distribution, particularly in situations where there is little control over the playback devices that will be used to read the discs. DVD-Video discs made using DVD-R media will be readable in the vast majority of DVD players, as well as those DVD-ROM drives equipped with the required decoding hardware. DVD-ROM discs produced using DVD-R equipment should be readable in all DVD-ROM drives.

Authoring Use

DVD-R for Authoring is the medium of choice for testing and development work, where developers must confirm operation on a range of target playback equipment before releasing a title to a replication facility for mass manufacturing. To avoid costly errors when a DVD title is being authored, DVD-R lets developers and testing firms produce discs that can than be run in standard playback equipment—either a DVD player or DVD-ROM drive. Any inaccuracies or problems with the playback can be detected and corrected before a disc get submitted for mass replication.

Presentations

Producing DVD discs for presentations, particularly presentations destined for portable DVD player playback, represents another ideal use for this medium. The interactive capabilities of the DVD-Video format make it possible to author presentations that are similar to full interactive multimedia applications.

Archiving

Certain archival applications, such as storing image data, audio material, motion pictures, and so on, favor the write-once characteristics of DVD-R. Archivists who want to record data and then ensure that it is not altered can rely on the properties of DVD-R media to protect their stored file contents. The estimated 100-year plus lifespan of the DVD-R discs also provides assurance that long-term archiving can be accomplished safely using this form of optical recording. For archival operations where data must be available for near-line access, DVD-ROM jukeboxes provide a means of storing extremely large quantities of data for convenient access. For example, a 100-disc DVD-ROM jukebox can handle close to a half-terabyte (470GB) of information.

For shorter term archiving and storage, the rewritable storage options provided by DVD-RAM, discussed in a following section, may provide a better alternative.

The DVD+R Alternative

DVD+R is another write-once format that has established a strong presence in the industry. The specification was forged by the DVD+RW Alliance. This alliance, composed of industry leaders such as HP, Philips, Ricoh, Sony, Yamaha, Dell, and Thomson, offers widespread playback compatbility with DVD-ROM drives and DVD players. With a baseline write speed of 2.4x, DVD+R can complete the writing of a DVD disc several minutes faster than a conventional 2x recorder. This format is essentially equivalent to the DVD-R General Use version.

DVD-RAM

The DVD-RAM format was the first of the rewritable DVD formats to reach the market. DVD-RAM drives are less expensive than DVD-R Authoring Use equipment, primarily because they use phase-change technology for storing data rather than the method used by DVD-R, which employs an organic dye to record laser impressions on disc.

The phase-change technology—which is an amalgam of the technologies used in magneto-optical cartridges, CD-RW, and PD devices—is also the reason that discs created on a DVD-RAM drive cannot be read in the typical DVD-ROM drive or DVD player. This significant drawback limits the utility of this particular storage method, but advances in player technology may overcome the limitation, much in the same way that early CD-RW discs could only be read in CD-RW players. Now, MultiRead-compatible players can handle CD-RW discs and CD-R discs with equal ease. Similar advances may make DVD-RAM discs more widely compatible.

Data patterns written to a DVD-RAM disc are recorded to a thin film that is sensitive to laser light. The DVD-RAM write laser strikes the film surface and converts the material from a crystalline state to an amorphous state. The reflectivity of these two different states is different enough that it serves to identify bit patterns. The disc surface can be "erased" by applying a different intensity laser burst. The energy from this burst converts the film from the amorphous state back to the original crystalline state.

DVD-RAM discs employ the same kind of modulation and error-correction codes that are used for DVD-Video and DVD-ROM. This characteristic should help ensure broadened future compatibility for this media type. Both single-sided and dual-sided media are available. Each side of a version 1 DVD-RAM disc has a capacity of 2.6GB. Version 2 DVD-RAM implementations doubled this to 4.7GB per side. In most regards, the physical discs have the same dimensions as DVD-R discs, but both single-sided and double-sided discs are housed in cartridges for use. Clock data is embedded in a wobble pattern integrated into the tracks, also offering a means by which address signals can be identified by the drive.

Uses for DVD-RAM

The primary appeal of DVD-RAM is inexpensive, flexible, abundant storage—this benefit comes at the loss of compatibility with the majority of players and drives in the market.

DVD-RAM is well suited to these kinds of uses:

- Short or long-term storage of press-quality images, digital audio files, digital video files, or other similar kinds of content requiring large amounts of storage space
- Local periodic backup and temporary archiving of network files, personal hard disks, organizational files, and so on
- Exchange of large-volume files between parties with similar equipment (compatible DVD-RAM drives)
- Nearline storage of mission-critical data that is important to an organization's operation, but cannot be concisely stored on the network
- Network-resident storage for periodic system backup or personal workstation archiving, through thin server technologies

As the DVD-RAM technologies matures and the adoption of the Super MultiRead standard makes it possible to exchange disc cartridges more freely, this format should continue to provide a flexible, inexpensive storage medium that provides a high degree of utility.

Re-Recordable Formats

DVD-RAM was first on the scene with a form of DVD storage that could be rewritten hundreds of times. Two additional formats, DVD-RW and DVD+RW, have emerged to offer a reusable medium that can be re-recorded up to 1000 times.

DVD-RW

Officially sanctioned by the DVD Forum, DVD-RW offers up to 4.7GB of storage on a cartridge-free medium that can be played back on many existing DVD players and drives. DVD-RW discs are frequently used for home and consumer applications, such as editing of home movies and storage of digital photographs and music files. The optical properties of a DVD-RW disc are similar to a commercial DVD-9 disc.

DVD+RW

Introduced as a faster, less expensive alternative to DVD-RW, the DVD+RW format enjoys widespread industry support and some additional features that improve start-and-stop recording. A technology known as lossless linking simplifies the mechanics of performing write operations and replacing 32KB data blocks during rewrites. Compatibility with current generation players and drives is nearly equal to DVD-RW, providing a good value and a high measure of data integrity through a defect management scheme that verifies the accuracy of data written to disc and read back.

Summary

The evolution of the DVD standards in many ways mirrors the course of CD-ROM and recordable CD development, but there are also some clear advantages to the way data storage has been handled on DVD. The format is far more flexible when it comes to embedded different data types onto disc, without the necessity for creating individual formats for each of the individual data types (as can be seen on CD-ROM with Video CDs, Photo CDs, CD-PROM, CD-ROM XA, and so on). The UDF standard also intelligently handles most of the key cross-platform issues, making DVDs less prone to the kind of file translation issues and file system concerns that have complicated CD-ROM delivery. The backwards compatibility with CD-ROMs and audio CDs, which is a requisite feature of most drives and players, also makes DVD the logical successor to the CDs as the ideal portable data storage medium. The newer recordable formats also make it exceptionally easy for developers, programmers, and filmmakers to generate content that can be played back on many different machines, including mainstream consumer DVD players. Delivering movies, multimedia content, games, high-resolution photographs, digital audio, and other content on DVD has become an extraordinarily useful vehicle for communicators in many different fields.

18

Distributing DVD Titles

Given the relative simplicity and easy availability of the tools required to make a DVD, this technology offers a wealth of opportunities for independent developers, filmmakers, small software firms, musicians, and others who prefer to do business on their own terms. While partnering with distributors and corporations with wide-reaching media connections can have financial benefits, it can also lead to frustration and financial disappointment in many cases. One of the chief benefits of DVD technology is that it can accommodate many forms of independent expression and, as demonstrated in the earlier case studies, it can form the basis of many types of unique business ventures.

At the same time, simply creating an innovative or inspiring DVD title doesn't automatically confer a passport to instant wealth or fame. As with any business, particularly those centered around an emerging technology, it generally takes persistence and diligence to attract an audience, nurture new customers, and gain recognition. This chapter addresses some of ways in which an independent DVD title producer can find an audience, generate customers, and build a sustainable business venture from digital content distributed on a quarter-ounce disc.

Reaching Audiences through the World Wide Web

The dot.com bust soured many entrepreneurs on the idea that a sound business plan could be built around the rapidly evolving, unpredictable World Wide Web. Consequently, an air of skepticism often prevails whenever the Web drifts into a business-oriented conversation. Rivers of spam threaten to bog down legitimate email communication across the Internet. A raft of well-funded Web-based ventures have died ignominious deaths, unable to generate revenues after the startup capital ran out. Many major corporations have reduced their Web presence to reduce their IT budgets. Web publishing ventures, such as *Slate* and *salon.com*, continually reorganize and regroup, seeking a viable business model and sustaining profitability.

Amidst these struggles and failures, many Web-based businesses have overcome the odds and achieved success. And, others are exploring new techniques for reaching audiences and gaining recognition. Democratic presidential contender Howard Dean has used the Web both for organizing support groups (through *Meetup.com*) and delivering campaign videos to prospective voters. Fund raising can also be impressively effective over the Internet. Dean's pre-primary victory in a poll conducted on *moveon.org*, a grassroots organization created to galvanize opposition to right-wing excesses, generated millions of dollars in campaign donations, adding credibility to Dean's campaign and placing him among the leaders of the Democratic contenders. The other candidates, after witnessing Dean's effective use of the Web, have attempted to use similar techniques to raise funds and communicate their messages. In the same manner as the Web can bring together people with common political aims, it can attract people with specialized interests who might be inclined to purchase your DVD title.

As a marketing arm of your business, the Web can be used to help promote your film or training DVD or documentary title. If you're so inclined, you can cut out the distributors and the middleman completely and sell your DVD titles directly to customers through a Web-based storefront. Web-based retail outlets such as Amazon.com also work with smaller producers to list DVD titles and capture orders through their well-refined order system; orders can then be fulfilled by the producer through whatever mechanism works most effectively for them.

The increasing prevalence of broadband Internet access in the home has created a surge of Web-based delivery of feature films through companies such as Movielink and CinemaNow. Cable television broadcast channels, always in the market for new content, have increasingly turned to independent filmmaking efforts, including short films, to reach specialized audiences. The Sundance Channel now airs features and shorts from their annual festival and the Independent Film Channel offers a broad selection of full-length feature films and short films, providing another avenue for independents to get their work seen.

Reversal: A Web Success Story

Reversal, a film created by Jimi Petulla for under $500,000, relates the story of a young wrestler and the obstacles he faces, both personal and competitive. Jimi based the tale on his own experiences and hoped to find a distributor for this work, but none of the distributors he contacted thought the film had wide-enough market appeal to be profitable. Discouraged, but determined, Jimi paid for some test screenings in Oklahoma and Texas theaters and found that *Reversal* was well received by audiences. Trying to finance a theatrical run himself, however, was clearly much too expensive. After some reflection, Jimi decided to seek an audience over the Web. He launched a Web site—*www.reversalthemovie.com*—and offered the movie in DVD format with a number of extras, including audio commentary tracks by Olympic gold-medal-winning wrestlers.

The combination proved unbeatable. In a 120-day period, Jimi sold about 13,000 copies of the title through his own Web site and Amazon.com. Jimi is actively working to extend the title availability to other chains and retail outlets. The title sales, Jimi thinks, have been largely spurred by growing awareness of the film within the tight-knit amateur wrestling community, as well as the power of the Internet to reach out to niche audiences.

Some of his audience was finding his Web site (and purchasing the DVD) using a search engine and some variation of the term *wrestling*. Others were finding the title on Amazon.com, which also has a sophisticated internal search engine. Well-placed links on other sites and word-of-mouth advertising also contributed to growing audience recognition of his work.

Growing a Business with the Web

Although the Web represents an indispensable tool for independent title producers, growing a business solely around Web sales can take incredible amounts of effort and ingenuity. When you stop and think about it, growing any kind of business in today's dynamic, fluctuating marketplace can take incredible amounts of effort and perseverance. If you make the Web a highly visible cornerstone of your business, either providing direct sales of titles and productions, or offering pointers to distributors or retail outlets where titles can be purchased, this Web presence can help drive your business.

Targeting Niche Markets

As an independent DVD title developer or producer, you probably don't have the funds to compete head-to-head with the well-established corporations that dominate the industry. On the other hand, you don't have their overhead to burden your operation—you can realize a significant profit on a small percentage of the sales that would be considered a success by one of the major players. While it may be difficult to release a general-purpose title and achieve success with it, you may find greater success by targeting your work to smaller audiences with niche interests. This lets you capitalize on market opportunities that the large-scale producers consider too small to be worth their efforts.

Often, the best ideas for DVD titles come from the deepest interests and passions of their creators, as demonstrated over and over in the case studies chapters of this book. Stefan Grossman turned his love for acoustic guitar music into a multi-faceted series of DVD titles, offering lessons, historical footage, and concerts starring exceptional guitarists. Chris DeHut shares his passion for creating finely crafted wood furniture through a DVD magazine, *Woodworking at Home*. Jeanette DePatie's desire to empower women and help them gain fitness resulted in a spirited workout DVD. If you'd like to tell a story or teach someone a skill using DVD as the medium, you can follow your interests and find a niche, no matter how narrow, and odds are that there is an audience out there who shares these interests. If you're looking for a starting project to learn the tools and techniques of making DVDs, consider the following ideas:

- Regional history and genealogy perennially interest many people, wherever they might live. Talk to the staff at your local historical society (almost every community has one) and see if they have any interest in helping create a documentary of significant events in the community's history. As a part of the project, they might be able to put you in contact with some of the older residents of the area so that you can capture oral histories from them. Ken Burns started a documentary dynasty with a single work about the building of the Brooklyn Bridge and established many of the techniques that he still uses to this day on projects.
- A DVD-ROM provides an excellent platform for a specialized reference that mirrors your interests. Produce a compendium of American motorcycles from 1910 to 1950. Design a resource guide for beginning kayakers that includes videos of paddling techniques and emergency procedures. Create a recipe book for vegans

with video sections that show different dishes being prepared. Develop an animated guide to alternative power and getting off the grid, including sections on solar panel placement and windvane installations. Produce a compilation of the most talented political cartoonists using Flash or QuickTime to communicate their ideas. By choosing a narrow segment of the overall market for a DVD or DVD-ROM title, you can reach an audience that is untapped by the mainstream title publishers.

- Resource databases are a popular and effective use of optical disc publishing. Find an appropriate vein to mine in this area and you can probably create an income-producing title. If you take advantage of abundant hyperlinks and indexing of terms, you can also make such a resource more usable than an equivalent printed directory or resource listing.
- Training and educational titles can appeal very successfully to niche markets. Pick an area where you have some expertise and share your knowledge with the rest of the world. If you enjoy historic architecture and know something about acquiring property and restoring it, produce a title on that subject. If you've mastered the latest non-linear editor and want to help others do the same, create a training DVD on the topic. If you've become skilled in a oriental martial arts program, you might have the material for a successful DVD title.

The range of material available to those working with digital content is virtually unlimited. You have the potential for building a sustainable business around supplying information that is needed in specialized areas or delivering entertainment to those with eclectic tastes.

Film Festivals and Animation Events

Getting your work seen, particularly in a creative venue such as independent filmmaking or animation, can take a grassroots effort, which may begin by combing the available resources in the film community. Tracking the upcoming festival events or animation competitions can offer one more means of getting your work seen and building an audience.

The bias for film over optical disc can be a major impediment to this process, particularly since the conversion costs for taking digital video to 35mm film are typically around $1500 to $2000 for a feature-length film. Fortunately, a growing number of film festivals and theaters are moving towards acceptance of digital forms of motion pictures, including the

DVD, and investing in projection systems that can handle digital media. Although the resolution of DVD images limits the scale of projection to smaller, more intimate theaters, higher-resolution shorts in Windows Media format can be effectively distributed on DVD-ROM and the next generation of high-definition DVD discs will make projected video much more commonplace.

Similarly, DVD-R technology makes it possible for an independent film company or video studio to create original masters and release them for replication in DVD-Video format. They can also create copies using the latest DVD duplicator technology, a far more cost-effective means of distributing movies than duplicating film. Software tools, such as *DVDit!* from Sonic Solutions, *VegasVideo+DVD* from Sony, and *DVD Studio Pro* from Apple bring impressive capabilities to the mainstream market, making it possible for anyone to turn digital video content into a published DVD work.

Opportunities abound for independent developers to produce successful works and realize a profit with unit sales in the hundreds and thousands, rather than the tens of thousands, as is the case with many commercial titles. The challenge, of course, is finding an audience in a market where many of the retail outlets are geared to large mainstream companies.

Screening DVD Movies

The barrier for independents who want to get their work screened in a theater is steep. While many filmmakers go the route of using digital video cameras and non-linear editing tools to create their move, they often have the final work transferred to 35mm film to be taken seriously in a market that is dominated by several large studios. The cinemas in most of the Western world are also thoroughly committed to the 35mm print format—a format that requires that completed films be physically distributed on heavy reels, often requiring several reels per film, as well as the accompanying hefty duplication costs.

This situation is changing rapidly as digital media continues to make inroads in the film industry. Landmark Theatres, a nationwide chain that specializes in both independent and foreign films, has equipped 185 screens in its 54 theaters with a digital media delivery system. Using a system developed by Digital Cinema Solutions (DCS), Landmark Theatres can present encoded Microsoft Windows Media 9 files delivered over a virtual private network (VPN), as well as DVD video content loaded directly into the system. A mobile cart equipped with a projector, computer server, and audio components features touch-screen controls, letting the projectionist control the presented material by organizing playlists of movies and trailers.

Landmark has a long history of nurturing independent filmmaking. Many of the theaters spanning 14 states reside in renovated historical buildings and these outlets showcase an eclectic assortment of films, including cult films, controversial releases, classic cinema works, and cutting-edge independent productions, such as *Monsoon Wedding, Bowling for Columbine, Run Lola Run*, and *Fast Runner*.

This approach has proven successful for the art-house chain, leading to growth and expansion in a number of metropolitan areas. The Kendall Square Cinema in Cambridge, Massachusetts boasts nine screens and enjoys ongoing support from the community, representing Landmark's most successful complex to date. Landmark expects to have the digital delivery system fully in place in each of their theaters by the end of 2003.

Many independent filmmakers see developments such as this as part of an irreversible trend leading towards a fully energized digital storytelling community. With the entry price of completing a digital film becoming lower every day and new outlets for digital works springing up everywhere, the looming barrier of theatrical distribution should be less imposing. This bodes well for those creative spirits who take on the challenge of creating a work that doesn't fit within the narrow confines of mainstream movie offerings.

Teaming up with a Distributor

Distributors offer risks and rewards—weighing the potential risks of signing on with a distributor can be one of the most difficult decisions that an independent DVD producer makes. Many distributors will do everything they can do in negotiating a contract with you to ensure that their risk is minimal and contingencies are in place so that they can reap a profit in a wide variety of circumstances. The methods they use to accomplish this can directly affect your bottom line. Does your distribution agreement specify a minimum price at which your title can be sold? If not, the distributor may arbitrarily drop the price, reducing your royalties, to move the title more quickly or to create bundled deals to attract greater sales.

Savvy DVD distributors know how to reach all of the potential commercial markets that your documentary or film project might reach, including cable television, theaters, rental services, and direct DVD sales on the Web and in stores. A new generation of distributors is rising in response to the popularity of the DVD, recognizing that the model for success in the industry is rapidly changing. In 2001, the sales of DVD titles surpassed all of the box office receipts for all of the films released to theaters. Clearly a new paradigm exists. Movie studios understand that many of their films will never make a profit on the screen, but they routinely count on VHS and DVD sales to generate a profit. If a distributor can effectively take your work and get it out in the channels that are most lucrative, including the multitude of new DVD markets, your chances of success are much greater.

Companies such as Plexifilm (*www.plexifilm.com*) and IndieDVD (*www.indiedvd.com*) have built businesses around the need for new distribution channels to service DVD creators who may not be aligned with the major movie studios.

Rental services, such as Web-based Netflix (*www.netflix.com*), also provide an outlet for DVDs that might be targeted towards niche markets. Netflix provides an extensive rating system for users, allowing them to rate large numbers of DVD titles. Knowledge management algorithms can then be applied to determine users' tastes and offer recommendations on movies that they may want to rent. This system helps open a market for much smaller areas of speciality. For example, Netflix knows that a certain segment of the audience enjoys documentaries, so they can announce new documentaries that appear in their listings to that select group of users. Netflix is much more inclined to purchase titles that appeal to smaller audiences because they can pinpoint those audiences so effectively. Distributors have found that they can sell titles to Netflix that might be overlooked or ignored by larger volume concerns.

Using the Internet as a Leveler

The Internet offers a variety of well-established channels for reaching potential customers, but if you want to achieve success, you need to find targeted methods to reach *your* audience. Possibilities include:

- Trading Web articles or other content in exchange for links from another site to yours. Larger Web sites are often desperate for fresh content to keep their site interesting to visitors, so you have a powerful bargaining chip at your disposal. Choose sites that match your specialty interest for the best results.

- Producing a monthly electronically published newsletter made available through the Internet, using a simple sign-up list. You can build the subscription list through postings in newsgroups or special-interest forums, sign-up forms on your Web site, or through opt-in mailing lists for people who have expressed an interest in receiving information about your topic of interest.
- Performing an opt-in mailing through a service such as YesMail (*www.yesmail.com*) to reach a targeted segment of the market. This can be an effective way to announce a new product release or stimulate interest in a title that has been available independently for some time.
- Using services, such as Overture's search term bidding, to selectively choose those terms that you want to use to identify your business on the Internet. Clever use of terms and skillful bidding can bring you to the top of the list in many select areas, as described in *Bidding on Search Terms* on page 273.
- Offering training in a specialized area and posting your training course on one of the Web sites that provides both the tools and the content storage. For example, if you have a product that facilitates 3D character animation, you could create a beginner's course on character animation and post it on *www.learn.com*. You not only gain attention from prospective customers, but you can achieve a level of credibility for your product or title.
- Starting a blog (Web log) to reach a niche audience, whether you have a DVD that extols the virtues of Greek cooking, the use of active solar energy techniques in northern climes, or the building of model railroads. A discreet link that leads to your title can generate sales. Blogger (*new.blogger.com/home.pyra*) offers what they call *Push-button Publishing for the People*. Other sites offer similar tools.

Imagination, creativity, and innovative communication can often make up for the sheer power of dollars when it comes to reaching an audience on the Internet.

Opening a Web Store to Sell DVDs

The entry requirements for constructing and maintaining a Web site equipped for ecommerce can be steep—too steep and too time consuming for many independent developers and small businesses. One solution is using services that provide the site, the commerce mechanisms, and the visibility to get you quickly online selling your titles or other products.

One popular service of this sort—Merchandizer (*www.merchandizer.com*)—offers a highly configurable storefront model. You can use their templates to display your goods or you can upload your own HTML pages at any level of complexity. Many added benefits to the Merchandizer approach make it easy for customers to interact with your site.

Other options for turnkey storefronts are available from different sources. Yahoo! Merchant Solutions (*smallbusiness.yahoo.com/merchant*) offers an electronic storefront that is easy to setup and use. By the time this book reaches the bookstores, many new options will no doubt be available. The Web storefront can be the focal point of your marketing effort and your primary sales channel. You can direct potential customers to it through email campaigns, links from other high-profile sites, search terms that are designated through bidding, newsgroup postings, and other techniques.

Targeted Press Releases

Press releases have long been an effective tool for large corporations. New product releases, merger announcements, patent filings, trade show announcements, and similar kinds of information often find their way into the mainstream press—including newspapers, trade magazines, radio, television—through targeted press releases. These are distributed to all of the listed press facilities and frequently incorporated into news pieces in a variety of formats.

You can compile a list of key press facilities and perform your own distribution of press releases. You can also rely on a service to do this for you. One company, Digital Works (*www.digitalworks.com*), offers inexpensive press release distribution to your choice of geographic locations, making it possible to achieve notice in a select market or location with your product news.

If you're not totally comfortable trying to write your own press releases, Digital Works, as well as other similar companies, provide specialists who can guide you through the process or write the release for you based on information that you provide.

Bidding on Search Terms

Fighting to get yourself noticed in the major search engines can be an enormously difficult task. The sheer volume of companies that are engaged in some form of Internet business has become overwhelming. An easier way might be to selectively place yourself in front of the audience that you want to reach by bidding on specific search terms.

Overture, which was recently acquired by Yahoo!, originated the Pay-for-Performance approach that can be a highly effective way for reaching very large audiences. Overture's distribution partners, who utilize the search service and display it to their customers, include CNN, MSN, Yahoo!, AOL Europe, Netscape, HP, and others. Overture provides a bidding system where you can choose the amount of money that you will pay for each click-through on a search term that links to your site. You only pay for click-throughs, not for views of the search term, so this can be an cost-effective way to steer customers to your site, especially when compared to banner ads which often register the times an ad has been "viewed" rather than clicked on.

The way to get the most mileage out of Overture is through clever search term selection. You want to choose those search terms that will attract the unique interests of those customers you want to reach, without picking terms that are too visible (such as sex, money, or MP3). The most visible terms are the ones that have the highest bids. The amount that you bid on a term determines where your placement is when the search engine displays listings for that term. This is why you need to steer away from the more common terms, since these become the ones involved in bidding wars, driving the prices upwards.

Let's say you've developed a DVD-ROM that provides a history of the use of pewter and its value as a commodity in colonial New England. You run a few test searches using the Overture engine and discover that there are no current bids on the term *pewter*. You place a minimum bid of $.10 on the term, and also select several other terms, including *colonial trading, New England history, oil lamps,* and so on. As part of the search term, you write a description that will appear when the search term is brought up. For example, you might want to attract interest with a phrase that reads: Learn about pewter use in Colonial New England. You then add your Web address. Each time someone clicks through to your site, after running a search on pewter, you pay Overture $.10. If you get 1000 visitors in a month, you pay $100.00. Keep in mind that because you are the high bidder on this term, each time someone searches for the term "pewter," your listing comes up at the top. For a small company competing with giants, being at the top of any search list can be a major competitive advantage.

Current bids on search terms typically range between \$.10 and \$2.00. At \$2.00 per click-through, this becomes a bit more expensive means of generating traffic, but, otherwise, it is probably the least expensive way you can significantly boost your Web site traffic in a very targeted way.

Through the selective use of different search terms, you can precisely identify interests in your prospective audience, which should generate some fairly good response rates as new visitors enter your site. The success that Overture has had partnering with other companies, many of them with an international presence, ensures that your selections of search terms will reach a very broad audience.

Summary

Niche markets offer great opportunities for small-scale title producers. Work within the niche that most interests you and then explore all viable outlets, as described in this chapter, to reach your target audience. The possibilities for small-scale DVD publishers are enormous and growing as increasing numbers of business and home users turn to DVDs for information and entertainment.

Glossary

1.33:1 Aspect ratio that applies to conventional television, Super-35mm film, and 16mm film. This corresponds with screen dimensions of 4 units horizontally for each 3 units vertically.

1.78:1 Aspect ratio that applies to high-definition television. This corresponds with screen dimensions of 16 units horizontally for each 9 units vertically.

2-3 pulldown A transfer technique used for converting film recorded at 24 frames per second to video. Alternating frames are captured as two video fields followed by three video fields.

24 fps Stands for 24 frames per second, the standard rate at which motion picture film is exposed and projected in most of North America and other countries. Europe and a number of other countries use a 25 fps rate for both television and project motion pictures.

24p This video format, referred to as 24 frames per second progressive scan, produces video content that approximates the characteristics of film.

4:1:1 The sampling rate used to record luminance to chrominance values for the DV format. The video brightness, called luminance, is sampled 4 times for each sample of the red and blue color information, called chrominance.

4:2:0 The sampling rate that applies to luminance and chrominance values used for the video on DVD.

4:2:2 The luminance to chrominance sampling rate associated with the D1 standard, offering higher resolution chroma samples, as compared to the DV format, which translates to better color reproduction.

8/16 modulation A technique used to store channel information on a DVD disc.

Absolute-time The time elapsed since the beginning of a recorded Red Book digital audio program; also known as A-time. A-time is calculated by reference to an internal clock that monitors elapsed time starting at the beginning of the innermost track.

AC3 The original name for the encoding scheme now officially termed Dolby Digital. The AC3 term still frequently appears in reference to descriptions of the process.

Access Time The time required to position the laser read head over a specified sector on a disc and begin retrieving the data.

Advanced SCSI Programmer's Interface (Usually shortened to ASPI. A set of ANSI-defined commands for application-level communication with SCSI host adapters. Most operating systems have ASPI drivers that are used to communicate with CD or DVD recorders included on a SCSI chain.

AGC Shortened form for automatic gain control. An electronic technique for boosting an incoming signal level to meet minimal acceptable recording strength.

Aliasing A form of image distortion caused by low sampling rates that creates jagged stairsteps within diagonal lines. Anti-aliasing techniques can smooth the appearance of these lines by intermixing pixels of varying color values along the jagged edges.

Analog to Digital Converter A hardware component that converts an analog waveform to a succession of digital values by sampling the waveform at periodic intervals. Often abbreviated to A/D converter or ADC.

Angles DVD-Video discs can potentially display up to 9 different camera angles, allowing the audience to switch the point of view through remote control.

Artifact A data abnormality that can appear in an audio or video file as the result of certain kinds of signal processing, including compression, data transfer errors, signal noise, or electrical interference.

Aspect Ratio The ratio of the horizontal size to the vertical size of a picture. In television, the aspect ratio is 4:3. In widescreen DVD, the aspect ratio is 16:9.

ATAPI Abbreviated form for Advanced Technology Attachment Packet Interface. ATAPI provides a layer of commands used to manage devices connected through an IDE bus, including CD-ROM and DVD-ROM drives.

Authoring The integration of the individual elements—sound, video, text, graphics—within a DVD production. The file structure and file encoding specified during the authoring process are used to burn a recordable DVD or as files to submit for replication, which may be transferred to another medium, such as Digital Linear Tape (DLT).

AVI Short for Audio Video Interleaved. A bitmapped audio/video format introduced by Microsoft that interleaves the audio and video content. Video capture tools and software on the PC platform often create raw AVI files, which can be easily edited, but they generally must be compressed further for distribution on disc or the Internet because of the large file sizes.

Autoplay Discs encoded with the autoplay option will begin immediate playback when inserted in a DVD player that supports this feature.

Bandwidth A measurement of the data-carrying capacity of a bus or other data transmission medium.

BCA Short for burst cutting area. An area located near the center of a DVD disc that is reserved for ID codes and manufacturing data. The BCA is imprinted as bar-code data.

Bidirectional frames A type of frame created by compression software that is inserted between intraframes and interframes. These frames contain averaged content from the neighboring frames and they are typically dropped during playback.

Bit Short for binary digit. A bit is the basic element representing digital data. Bits are combined into groups to form bytes (8 bits), words (16 bits), and double-words (32 bits).

Bi-refringence A term applied to the refraction of a beam of light in two different directions. This phenomenon occurs as an undesirable aspect of the compact disc manufacturing process resulting from residual stresses in the polycarbonate substrate introduced during injection molding. Excessive bi-refringence results in laser read errors.

Bit Error Rate Often shortened to BER; indicates the number of bit errors that occur in proportion to the number of correctly processed bits.

Bits Per Pixel Sometimes referred to as color depth, the number of bits per pixel defines the maximum color variations available for each pixel that appears onscreen. 8-bit color allows 256 individual colors. 16-bit color allows 65,536 colors, and so on.

Block Error Rate Often shortened to BLER; indicates the number of blocks in which an error was detected in proportion to the overall number of blocks processed.

Book A The basic specification for the physical format that applies to DVD discs; this specification forms the basis for the DVD-ROM.

Book B The specification that defines the format of DVD-Video discs.

Book C The specification that defines the format of DVD-Audio discs.

Book D The specification that defines the format of DVD-R (write-once).

Book E The specification that defines the format of rewritable versions of DVD.

Buffer A temporary storage area used to compensate for differences in the data transfer rates of two devices. The buffer holds a quantity of data that ensures a continuous flow to the faster device while the slower device works to keep the buffer full.

Camcorder A type of video camera that records a series of images and stores them in electronic format.

CAV Shortened form for constant angular velocity. In this type of data storage system, the rotation speed of the disc is kept constant as the read/write head is positioned over different points on the disc.

CCD Shortened form for Charge-Coupled Device. A light-sensitive integrated circuit that captures image data and converts it to electronic signals. In a video camera, the size of the CCD, the number of CCDs used, and the pixels available for image capture determine the relative quality of video.

Cell As applies to DVD-Video, a single unit of video content that can vary in length from less than a second to several hours. This structure allows video content to be grouped in various ways for interactive playback.

Channel In audio terms, a division of the audio content that is typically directed to one speaker. For example, stereo signals include two channels of audio content.

Challenge Key Part of the encryption process used in DVD-ROM content presentation, the challenge key authenticates an exchange between the drive and host computer.

Channel A color component of a computer graphic image, such as the RGB component or the alpha component.

Chapter A basic division of the content on a DVD-Video disc. The technical term, part of title (PTT) is sometimes used to refer to a chapter. A chapter consists of a scene or particular video segment that is defined during authoring.

Chroma Key An effect available in many video applications that allows a specific color (usually blue or green) to be removed from the video and additional video added to these regions.

Chrominance The values that represent the color component of a video image. These values are combined with the intensity data, known as luminance, to define the appearance of the image.

Closed Captions Textual content that is embedded in the MPEG-2 stream of a DVD. This data can be displayed by users during playback through selections on the remote control unit or television console. The text is transferred to the television during the vertical blanking interval of the video signal.

CLV Shortened form for constant linear velocity. Refers to the varying of the disc rotation speed to ensure that the laser read head encounters data at a constant linear rate (1.2 to 1.4 meters per second). Disc rotation speeds increase for inner tracks and decrease for outer tracks to maintain this constancy.

Codec Acronym for coder/decoder. A signal processing algorithm responsible for compressing (coding or encoding) and decompressing (decoding) an audio or video file.

Component Video A means of representing video as three individual components, such as RGB or YUV.

Composite Video A standard video signal in which the red, green, and blue components are combined with a timing (synchronization, or sync) signal. NTSC uses composite video. The process of creating composite video generates color artifacts, which makes it unsuitable for professional video or as source material for compression.

Compression A process by which a file is reduced in size by removing or encoding extraneous information. A lossy compression standard, such as JPEG, cannot regenerate the original file content to the same degree of accuracy. A lossless compression standard, such as GIF, can restore all the original information.

Control Area In DVD terms, a portion of the Lead-In area that contains a single ECC block, containing key disc data, repeated 16 times.

CSS Shortened form for Content Scrambling System. CSS offers digital copy protection for DVD-Video discs by mixing up audio and video data on the disc through encrypted keys. This process typically takes place during the glass mastering phase of disc replication.

Cue Sheet A sequence of tracks that designates the points in a film or video where sound effects, music, or other effects are to occur. The cue sheet becomes a guide during the editing process.

DAT Shortened form of Digital Audio Tape. A storage format for audio information on 4mm tape cartridges. Sony originated the techniques for sampling audio data and converting to a digital framework while designing their original DAT drives. These sampling and encoding techniques were than adapted to compact disc.

Data Area In DVD terms, the physical region residing between the Lead-In and Lead-Out areas where the actual data content of the disc appears.

Data Capture Those techniques for converting data from non-computer formats (video tape, photographs, line drawings, pages of text) into a digital form that can be processed by a computer.

Data Compression A process for reducing the storage space needed for data by compressing repeating information, such as a string of blank spaces in a text file or a block of pixels the same color in a graphic image.

Data Conversion Transferring information from one type of storage format into another. For example, a GIF graphic file can be converted into a BMP file, or a Word for Windows file can be converted into ASCII text.

DCT Shortened form for Discrete Cosine Transform. DCT is a video compression technique that is employed for DV format content.

De-esser An audio filtering technique designed to remove sibilant sounds from vocal material.

Deinterlacing A technique for removing the artifacts that are produced when interlaced video is captured and processed.

Digital Video The common video format that applies to DVCAM, miniDV, and DVCPRO, used for both video production and editing. The large file sizes make it necessary to apply further compression before delivery on disc or Internet streaming.

Digitize To convert from an analog source to digital form. For example, you can digitize an analog waveform to create a digital representation consisting of a string of binary values.

Digitizer A device used to convert video signals into a digital format suitable for display on a computer screen. Digitizers typically take several seconds to convert an image, and therefore require that the video picture be perfectly still.

Direct Memory Access Shortened to DMA. A technique for providing rapid transfer of data between a storage device and computer memory without requiring processor intervention.

Disc Array A collection of hard disks that be accessed as a single drive to increase the performance or provide a measure of redundancy.

Disc-at-Once A method of single-session recording in which the entire CD is written, from start to finish, without stopping the laser. The table of contents and Lead-in area are written initially, so this data must be compiled by the CD recorder software before the operation begins. The Disc-at-Once capability is necessary in order to prepare CD masters to submit to a replicator.

Disc Description Protocol Shortened to DDP. A protocol used to describe a compact disc at the sector level. DDP is used to ensure reliable and consistent mastering and is often the preferred format at replication facilities.

Disc Key The necessary value that must be provided to descramble the title key contained on a DVD-Video disc.

DLT Shortened form for Digital Linear Tape. A data storage system that uses tape-based serpentine recording to cartridges that offer 40GB capacities. DLT is one of the preferred methods for submitting files for DVD replication.

Dolby Digital A technique developed by Dolby Laboratories for encoding audio files using perceptual algorithms. Most DVD-Video discs utilize Dolby Digital for the stored audio content.

Dolby Surround A technique devised by Dolby Laboratories for encoding surround sound audio channels so that they can be presented to a stereo system.

Drop Frame Timecode A method of expressing a timecode that adjust for the 29.97 fps rate used by NTSC video. Compensates for the difference in frame counts by dropping two frames of timecode each minute and then skipping the drop every 10 minutes. Drop frame timecode helps in a number of applications, such as broadcast television, where the recorded timecode in a program needs to correspond with real time.

DV Shortened form for Digital Video. Typically applies to the standard developed jointly by Sony and JVC for their version of the digital videocassette.

DV25 A fixed rate of compression, 25 Megabits per second, that applies to the DV format. DV25 is the most commonly employed compression technique for DV.

DVD Shortened form for Digital Video Disc or Digital Versatile Disc. Applies to the audio-visual optical storage medium based on 120-millimeter discs.

DVD-Audio A storage medium for digital audio that offers improved bit depths and sampling rates in comparison to the CD.

DVD-R A form of writable DVD that employs a dye sublimation technique to record data. Discs in this format can only be written once.

DVD-RAM A form of writable DVD that uses phase-change technology, similar to the CD-RW format, to allow multiple write operation and erasures on the disc surface. DVD-RAM media is contained in a cartridge.

DVD-ROM The original DVD format that encompasses both DVD-Video and DVD-ROM (Read-Only Memory). These discs can only be read—they cannot be recorded. The standard includes a number of different data types and a file system: UDF.

DVD-Video A storage medium designed for playback of audio and video content on set-top devices known as DVD players. Content stored on DVD-Video discs can include MPEG video, Dolby Digital audio, MPEG audio, as well as other formats.

Dye Sublimation The method of recording data by focusing pulses from a laser beam onto an organic dye material, which records marks that can be read as pits.

Dynamic Range An audio term that distinguishes the difference between the softest parts and the highest volume parts of an audio signal.

Electroforming The technique employed to produce a metal master disc from a glass master. An electroplating process coats the glass master, retaining the embedded pits. The resulting disc, sometimes called the father, can then be used to produce a mother and a series of stampers for actual disc pressing.

Electronic Publishing Converting a print version of a document or presentation to a digital representation. An electronically published document can usually be used in diverse ways, such as distributed it on DVD-ROM or displaying it on the Web.

Elementary Stream A coded bitstream, composed of groups of packets, that is used to transfer audio or video content.

Encoder A device that converts and RGB video signal into a composite video signal.

Enhanced IDE Shortened to EIDE. A term that is commonly applied to a number of extended specifications based on the original IDE (Integrated Device Electronics) specification.

Encryption A process by which data is secured by encoding it in a form that it cannot be read without being decrypted.

Error Correction Code Shortened to ECC. A means of representing the information in a string of data so that the data can be reconstructed if errors occur during transfer.

Error Detection Code Shortened to EDC. A technique for using 32 bits in each CD-ROM sector to ensure the integrity of the transferred data. The EDC can be used to detect errors that occur during transfer; the ECC can be used to make corrections to the flawed data.

Fast SCSI An extension of the original SCSI-1 bus standard, Fast SCSI (also called SCSI-2) supports both 16-bit or 32-bit data transfers at rates up to 10MB per second.

Field The even- or odd-numbered scan lines that constitute one-half of a television picture.

File System An organization of the logical elements of a collection of data, such as the files and directories, so that they can be located on the physical media, segmented by sectors.

FireWire A high-speed data transfer method that is commonly used for video and audio content (such as transferring the output of digital camcorders). The IEEE 1394 specification formalizes the FireWire standard introduced by Apple. FireWire connections to disc recorders and digital camcorder have become very popular.

Fragmentation The scattering of the individual parts of files throughout the surface of a hard disk. Accessing fragmented files takes longer, which is why disk optimization—to eliminate fragmentation—is generally recommended before beginning compact disc recording.

Frame The complete television picture, consisting of two interlaced fields (see interlaced video).

Frame Grabber A device that digitizes video at real-time rates.

Frame Rate The number of image frames that occur per second in displayed video or film. The film rate in the U.S. is 24 fps. NTSC frame rates, that prevail in much of North America, are 29.97 fps. PAL frame rates, in much of Europe, are set to 25 fps.

Genlock A technique for mixing two or more video signals and ensuring that they remain in step. Combining video signals without genlock results in distortion.

Gigabyte Shortened to GB. A measurement of computer data consisting of 1,024 megabytes, or 1,073,741,824 bytes.

Glass Master The initial recording medium in disc replication. The glass master is treated with a photo-sensitive coating and the data is recorded using a laser beam. Treating the exposed glass master creates the pattern of pits.

GOP Shortened form for Group of Pictures. An MPEG video sequence typically consists of one or more I-frames followed by a series of P and B pictures. A GOP provides a unit of access for MPEG production.

Group A division that applies to DVD-Audio discs; each disc can contain up to 9 groups and each group can contain up to 99 tracks.

Huffman Coding A form of compression that assigns variable-length codes to precise value sets. A variation of Huffman coding, used with MPEG, relies on fixed code tables. This method of compression is lossless.

Incident Light Meter A device that can determine the amount of light striking an object and display a reading, in analog or digital format, of that measurement.

Injection Molding A manufacturing technique used during replication of compact discs. Molten plastic is injected into a mold and cooled to produce the disc. A stamper embeds the data pattern onto the disc surface as a part of this process.

Interlaced Video The process by which two separate video fields form a television picture. In the NTSC format, the field consisting of the odd-numbered lines is drawn first, followed 1/60 second later by the field with the even-numbered lines. Although the two fields don't actually appear at the same time, the human brain interprets them as a single frame lasting 1/30 seconds.

Interleaving The alternate placement of audio and video data with computer data to permit faster access and closer synchronization of sound to onscreen displays.

JPEG standard A compression scheme formulated by the Joint Photographic Experts Group that employs a DCT algorithm.

Jukebox A DVD-ROM player that handles more than one disc. The jukebox contains a mechanism, similar to the music jukeboxes that handled 45rpm records, that locates and mounts a particular compact disc for reading. Jukeboxes typically hold between 6 to 100 compact discs.

Lacquer Coating A protectant that is used to seal the surface of an optical disc after the data pattern has been imprinted.

Laserdisc An optical disc that stores analog data in FM format for playback of movies and interactive multimedia content. These discs range in size from 8 inches to 12 inches. Laserdiscs are still occasionally employed in video training applications and for high-quality movie viewing, although DVD-Video now offers superior video characteristics.

Latency The inherent delay experienced by the laser read head when locating specified data.

Lavaliere Microphone A small sensitive microphone generally used for vocal work. A lavaliere microphone usually attaches to the clothing somewhere on the upper chest. Audio can be relayed to a camera either wirelessly or through a microphone cable.

Layered Error Correction Code (LECC) A technique for correcting errors that cannot be handled by the CIRC. The LECC reprocesses detected errors and attempts to perform the correction using the EDC and ECC values.

Lead-In Area An area on a disc that precedes the data region, generally about 1.2mm wide. The lead-in area on a DVD contains information about the type of disc and the disc contents.

Lead-Out Area The area that follows the program area on a DVD. The lead-out area is at least 1.0mm wide.

Letterbox A technique for displaying films in their original format by placing black matte regions on the top and bottom of the image area. This allows a widescreen image to be placed on a standard TV with its 4:3 aspect ratio. DVD-Video players can typically apply this feature automatically.

Locked Audio A method of recording audio content in precise synchronization with recorded video.

Luminance A measurement for the intensity or brightness of a video signal, often represented by the letter Y.

Lux A unit of light intensity.

Macrovision A company engaged in producing copy protection schemes for VHS tapes, CDs, and DVDs.

Mastering The physical act of etching data pits into the photoresistant layer of a glass master in preparation for creating a metal stamper.

Matrix Encoding A technique that allows a number of surround sound audio channels to be presented to a conventional stereo system. A mathematical model is used to extract the appropriate audio information and deliver it in the proper format for stereo.

Matte A portion of the screen which is blackened or otherwise covered to change the aspect ratio of an image being presented on a monitor or television screen. A matte is typically applied to the top and bottom of the screen when using the letterbox format in DVD-Video playback.

Megabyte (MB) One million bytes of computer information. One byte is 8 bits. One bit is a single 1 or a 0, the basis of computer binary arithmetic. Eight bits grouped together form a byte which counts (in binary) up to 2 to the power of 8 = 256.

MJPEG Shortened form for Motion JPEG. A highly scalable video compression format which is utilized by many types of video editing equipment, such as the Media 100 and Avid suites.

Moire Pattern A convergence of lines in a video image that creates a distracting, undesireable pattern.

Mosquitoes A reference to a form of distortion that occurs after video compression. Mosquitoes appear as patterns of fuzzy dots that often appear around sharp transitions in the image. This effect is more formally known as the Gibbs Effect.

MPEG A standard formulated by the Motion Picture Experts Group to perform high compression of video data for reproduction on a variety of media, including CD-ROM and DVD-ROM. The processor intensive algorithms used for coding and decoding the video data work best when coupled with specialized hardware designed to accelerate this process. MPEG-1 and MPEG-2 have proven popular outside of the United States for distributing films and videos on CD-ROM. MPEG-2 is now the standard interlaced video format used for DVD-Video.

MPEG Audio A compression technique that uses perceptual encoding for storage and delivery. MPEG-2 audio format includes individual multichannel audio content.

Multiangle An option used in DVD-Video productions that allows the user to select one of several different viewing angles for the video content.

Multichannel An audio technique that employs separate channels of audio content that are directed to different speakers, often speakers configured to optimize a surround sound effect.

Multilanguage A capability of DVD-Video in which the production can contain indivdual sound tracks or subtitles that are presented in a number of different languages.

MultiRead A standard that was developed to allow standard CD-ROM drives and other playback devices, including DVD-ROM drives, to read various CD formats, including CD-RW discs. A similar standard, termed Super MultiRead, has been developed to provide similar compatibility for the different forms of DVDs.

Negative Cut The process of editing a negative by physically cutting and reassembling the film according to a cut list that is created by specifying timecodes from an edited work print. Even in the age of automation, the negative cut is generally a manual process.

Noise An unwanted, meaningless component of a signal that is generated by the recording process or some other aspect of signal processing. Digital audio and video technologies are relatively free from noise in comparison with analog methods of signal processing.

Non-Drop Frame Timecode A method of expressing a video timecode in which a full thirty frames per second is used.

NTSC The television standard currently in use in North America and Japan, an acronym for the National Television System Committee that created the standard. Detractors of the now-aging standard sometimes refer to it as "never the same color".

Optical Disc Any of the family of discs that relies on light, usually a laser beam, to read the data recorded on the disc.

Optical Head The term sometimes applied to the laser read mechanism of an optical disc drive or recorder.

OSTA Shortened form for Optical Storage Technology Association. A trade organization active in all forms of optical storage that helps craft standards and further the understanding of the underlying technologies.

OTP Shortened form for opposite track path. One variation of the data pattern used in a two-layer DVD disc in which the data begins near the center of the disc on the first layer and progresses to the second layer travelling from the outer edge to the inner. This technique is typically applied to very long programs designed for continuous playback.

Overlay The process by which computer graphics are combined with video, to add titles and animation to a scene, for example.

Pack A unit of MPEG packets contained in a DVD-Video playback stream. Packs consist of the contents of a DVD sector, containing 2048 bytes.

Packet In DVD-Video terms, a unit of storage that consists of a sequence of data bytes associated with an elementary stream. Packets are clustered into packs within the storage system.

PAL/SECAM These are foreign counterparts to the NTSC video standard. PAL (Phase Alternating Line) is primarily used in Western Europe while SECAM (Sequence de Couleurs avec Memoire) is used in France, Eastern Europe and the Soviet Union.

PCI Shortened form for Presentation Control Information. A data stream used in DVD-Video that carries timing details and other data, such as the selection information, aspect ratio, and so on.

Perceptual Coding A technique for compressing data that relies on human perceptions to determine what information to remove. Data that is least likely to be noticed can be extracted, while data that is more readily perceived is maintained. This compression technique is lossy.

PES Shortened form of packetized elementary stream. A low-level stream composed of MPEG packets. A PES might include audio or video content as an elementary stream.

Phase-Change A recording technique used for CD-RW and DVD-RAM that uses a laser beam to alternately change the state of the recording material. The recording layer is heated at one temperature to bring it to a crystalline state (when erasing) and another temperature to return it to an amorphous state (when recording a pit).

Pit A tiny impression in the surface of an optical disc that shifts the phase of reflected light from a laser beam. Pits are surrounded by lands; data transitions are recorded when a change in the data surface from land to pit or from pit to land is detected. Pits form the patterns of data that compose the information carrying layer of a CD-ROM or DVD. The size of a pit is approximately 0.5 by 2.0 microns.

Pitch The radial distance separating tracks on a compact disc (typically 1.6 microns)

Pixel Short for picture element, equals one dot on a computer display. A standard size for multimedia display is 640 pixels across and 480 down.

Plug and Play A technique for automatic installation of hardware and software that supports dynamic self-configuration. I/O ports, IRQ lines, and DMA channels are automatically assigned to non-conflicting values. Windows NT/2000/XP is one operating system designed around Plug and Play principles.

Polycarbonate The plastic-based material composing the substrate of a compact disc upon which the reflective metal surface is layered.

Premastering The preparation of digital data for recording to an appropriate DVD format. Premastering includes the creation of DVD control and navigation data, the combining of data streams, the calculation of error-correction codes, and channel modulation operations. Encoding is also an element of premastering.

Program Area The area on a CD or DVD disc containing the actual audio and video data. This is the largest single region on a disc.

Progressive Scan A alternative technique for scanning a television display that differs from an interlaced scan. The frame rate is increased from 30 Hz to 60 Hz and all the lines of the display are presented sequentially (rather than alternating odd and even). In PAL implementations, the frame rate is increased from 25 Hz to 50 Hz.

QuickTime An architecture designed by Apple Computer for presenting different types of digital media, including video, audio, and still images.

Raster Scan The pattern in which a video screen is scanned, usually from the upper-left corner to the bottom right.

Red Book The original compact disc specification designed for the storage and playback of audio information. Subsequent CD-ROM standards are based on the Red Book standard.

Redundancy The addition of data that makes error checking and correction possible when associated with a set of primary data. Redundancy techniques allow the reconstruction of information when a portion of the data is erroneously transferred.

Reed-Solomon Code An error correction code that uses algebraic principles to compensate for the types of errors that commonly occur with optical disc data. On a DVD, rows and columns using the Reed-Solomon code are contained in a two-dimensional lattice known as the Reed-Solomon product code. This form of error correction helps compensate for scratches or defective regions on a disc.

Refresh Rate The number of times per second that a video screen is repainted. NTSC video is shown at 30 frames per second while PAL and SECAM are displayed at 25 frames per second.

Replication The physical process of creating multiple copies of compact discs from a stamper using injection molding techniques.

Retrieval The act of locating a specified piece of data, usually used in respect to accessing information in databases.

RGB NTSC This is NTSC color-composite video decoded into its red, green, blue, and sync components. Broadcast-quality systems tend to use this format because the image quality is higher.

S-Video A form of NTSC video in which the chrominance (color) and luminance (brightness) signals are separated. S-Video produces slightly higher-quality images than color-composite NTSC.

Sampling Rate The rate at which a continuous waveform is measured as it is converted into a series of digital samples. Compact disc audio is conventionally sampled at the rate of 44,100 times per second. Generally, the higher the sampling rate, the more precise the audio or video capture.

Scan Converter A device that converts between video formats such as NTSC and PAL. These devices are usually very expensive.

Scan Lines The individual lines in a video picture. They are composed of pixels in the computer world and analog signals in the video world. A set of scan lines makes up a field.

Sector A unit of storage on a DVD disc that contains 38,688 bits of channel data and 2,048 bytes of user data. A sector is the smallest addressable unit on a disc.

Seek The physical operation associated with positioning the read/write head of a storage device in the proper location to read or write a particular piece of data. For CD-ROM applications, the seek operation generally requires also varying the rotational speed of the disc in relation to the radial position of the laser read head.

Seek Error The inability to identify and locate required data on a disc. Seek errors can be caused by surface irregularities, improper focusing of the laser read mechanism, or shock and vibration.

Servo Mechanism A motorized mechanical assembly that controls precise movements in response to voltage signals and a feedback circuit. Optical disc drives and recorders use servo mechanisms to control the positioning of the laser head over the disc surface.

SGML Shortened form of Standard Generalized Markup Language. An elaborate set of definitions specifying the formatting of documents intended for electronic distribution. SGML generally makes electronic publications accessible on a number of different computer platforms.

Small Computer System Interface Abbreviated to SCSI. An interface standard used for computer peripherals, such as scanners, hard disk drives, and disc recorders, that supports high-speed transfers over a common bus for up to seven devices per host adapter.

SMPTE Time Code A method of indicating precise time by means of recording signals used to synchronize various events. This technique was devised by the Society of Motion Picture Television Engineers and breaks down events to hours, minutes, seconds, and frames.

Stamper The metal plate created by an electroforming process from a mother, used in the injection molding of compact discs.

Storyboard A means of visually depicting the flow of events in a film or video by creating drawings or images, several panels to a page, illustrating the key scenes. Storyboards are used as communication tools to help production teams coordinate activities.

Subpictures Overlays consisting of image bitmaps that are used to present certain types of information on a DVD, such as subtitles, menu highlighting, and captions.

Substrate The primary material that give weight and form to a compact disc. The polycarbonate substrate of a disc contains data that has been stamped or deposited.

Sync An electronic metronome used to keep video signals in step with one another. In the NTSC color-composite standard, sync is combined with the red, green, and blue signals. In RGB systems, the sync signal may exist separately or be combined with the green signal.

Sync Point A point in a film or video where visual and audio elements occur together.

Telecine A process by which film content can be transferred to video. The telecine equipment projects the film frames in a specific sequence, referred to as 2-3 pulldown, to compensate for the lower frame rate of film.

Throughput The sum quantity of data that can be moved through a given data channel. Throughput measures the efficiency and capacity of data transfer within a system.

Timeline A means of organizing events while performing non-linear editing by assembling the elements of a project in respect to points of time in a graphically displayed sequence.

Track A unit of audiovisual information. A track on a DVD-Video may consist of a single track of video, as many as 8 tracks of audio, and a maximum of 32 tracks of subpictures.

Transfer Rate A measure of the speed at which data can be moved from one point to another. Transfer rates are generally expressed in terms of kilobytes per second.

Transition The type of change that identifies the movement between two video segments. For example, a wipe progressively replaces one segment with another segment by means of a moving pattern.

Treatment A concise summary of a film script used to solicit funding or interest in a project.

Two-Pass VBR A method employed by codecs to improve the compression of video content. Two passes are performed on each clip. The first pass analyzes the individual areas of the clip and determines which areas have high motion and which have low. The second pass actually performs the encoding, devoting more bits to high motion areas and fewer to low motion. Two-pass VBR creates more compact video files with higher quality than single-pass approaches.

UDF Shortened form for Universal Disc Format. A random-access file system devised by OSTA for use on a variety of optical media, including CD-RW and DVD-ROM.

UDF Bridge A form of UDF that also includes backwards compatibility to earlier devices that rely on ISO 9660.

UltraWide SCSI A high-speed implementation of the SCSI standard that can achieve transfer rates up to 40MB per second using a 16-bit data path.

Underscan A capability of some professional monitors that presents an entire video image in the display area. This feature compensates for the overscan on a conventional television monitor that results in the edges of the video overflowing the display region.

Unlocked Audio The method that applies to DV audio recording by which the audio data is not precisely linked to the video content. This can result in audio data that is as much as one-third of a frame out of sync with the captured video.

Variable Bit Rate Abbreviated to VBR. A technique for automatically adjusted the sampling rate when encoding video to compensate for the complexity of the content. Fast motion sequences and segments with intricate patterns or a multitude of colors require a higher VBR for accurate representation than do slower or less complex sequences.

Video Manager Abbreviated to VMG. Refers to the disc menu. Sometimes referred to as the title selection menu.

Video Title Set Abbreviated to VTS. An element of a DVD-Video disc consisting of one to ten files that contain the contents associated with a title.

Virtual Image A collection of files arranged in a particular order for either recording directly to compact disc or for creating an ISO 9660 image on hard disk in preparation for recording.

VOB Shortened form for Video Object. A file contained on a DVD-Video disc that contains multiplexed video, audio, and navigation information.

VU Shortened form for Volume Unit. A unit of measurment that applies to the audio level of a signal determined by averaging. Non-linear editors often feature software versions of VU meters that can be used to evaluate the volume of audio segments.

Watermark A method for positively identifying the owner or creator of a video segment by overlaying a still image, such as a logo, in a portion of a frame. Watermarks can identify videos that have been illegally duplicated or pirated.

Waveform Audio A sound file that contains a representation of an analog signal in digital form. Depending on the sample rate at which the waveform was digitized, whether stereo is enabled, and the number of bits per sample, the sound file can range less than 1MB per minute of recording to more than 10MB. The larger the file size, the better the quality. This format, usually represented with a .WAV file extension, is popular in Microsoft Windows.

White Balance A means of adjusting the light response of a video camera to compensate for existing lighting conditions. Properly setting the white balance adjustment on a camera enables the color values to be recorded correctly.

Wide SCSI An implementation of the SCSI standard using 16-bit data paths that can reach data transfer rates of 20MB per second.

Widescreen The presentation of a video image that exceeds the conventional 4:3 aspect ratio. In respect to DVD, a widescreen image typically has an aspect ration of 16:9.

Windows Media A flexible, compact storage format for audio and video data. Windows Media 9 files are used for a number of high-resolution applications, including high-definition digital cinema use.

Wizard A software utility designed to simplify the use of a particular task, such as preparing the files for encoding to streaming audio or video formats.

YUV A technique for encoding color images in which the video is dividing into luminance (Y) and two chrominance values (UV). Because the human eye can perceive intensity variations more keenly than color variations, luminance is sampled at full bandwidth while chrominance inforamtion is captured at a lesser bandwidth. Color values are stored as a measure of the red component (U) and the blue component (V). The difference between the red and blue values can be used to calculate the green component.

Zoom Range A distance that indicates the span of a wide angle lens setting as compared to the maximum telephoto setting.

Index